Managing
OCD with CBT
FOR
DUMMIES®
A Wiley Brand

by Katie d'Ath and Rob Willson

FOR
DUMMIES®
A Wiley Brand

Managing OCD with CBT For Dummies®

Published by: **John Wiley & Sons, Ltd.**, The Atrium, Southern Gate, Chichester, www.wiley.com

This edition first published 2016

© 2016 John Wiley & Sons, Ltd, Chichester, West Sussex.

Registered office

John Wiley & Sons Ltd, The Atrium, Southern Gate, Chichester, West Sussex, PO19 8SQ, United Kingdom

For details of our global editorial offices, for customer services and for information about how to apply for permission to reuse the copyright material in this book please see our website at www.wiley.com.

All rights reserved. No part of this publication may be reproduced, stored in a retrieval system, or transmitted, in any form or by any means, electronic, mechanical, photocopying, recording or otherwise, except as permitted by the UK Copyright, Designs and Patents Act 1988, without the prior permission of the publisher.

Wiley publishes in a variety of print and electronic formats and by print-on-demand. Some material included with standard print versions of this book may not be included in e-books or in print-on-demand. If this book refers to media such as a CD or DVD that is not included in the version you purchased, you may download this material at http://booksupport.wiley.com. For more information about Wiley products, visit www.wiley.com.

Designations used by companies to distinguish their products are often claimed as trademarks. All brand names and product names used in this book are trade names, service marks, trademarks or registered trademarks of their respective owners. The publisher is not associated with any product or vendor mentioned in this book.

LIMIT OF LIABILITY/DISCLAIMER OF WARRANTY: WHILE THE PUBLISHER AND AUTHOR HAVE USED THEIR BEST EFFORTS IN PREPARING THIS BOOK, THEY MAKE NO REPRESENTATIONS OR WARRANTIES WITH RESPECT TO THE ACCURACY OR COMPLETENESS OF THE CONTENTS OF THIS BOOK AND SPECIFICALLY DISCLAIM ANY IMPLIED WARRANTIES OF MERCHANTABILITY OR FITNESS FOR A PARTICULAR PURPOSE. IT IS SOLD ON THE UNDERSTANDING THAT THE PUBLISHER IS NOT ENGAGED IN RENDERING PROFESSIONAL SERVICES AND NEITHER THE PUBLISHER NOR THE AUTHOR SHALL BE LIABLE FOR DAMAGES ARISING HEREFROM. IF PRO-FESSIONAL ADVICE OR OTHER EXPERT ASSISTANCE IS REQUIRED, THE SERVICES OF A COMPETENT PROFESSIONAL SHOULD BE SOUGHT.

For general information on our other products and services, please contact our Customer Care Department within the U.S. at 877-762-2974, outside the U.S. at (001) 317-572-3993, or fax 317-572-4002. For technical support, please visit www.wiley.com/techsupport.

A catalogue record for this book is available from the British Library.

Library of Congress Control Number: 2016930758

ISBN 978-1-119-07414-4 (pbk); ISBN 978-1-119-07415-1 (ebk); ISBN 978-1-119-07416-8 (ebk)

Printed in Great Britain by TJ International, Padstow, Cornwall.

10 9 8 7 6 5 4 3 2 1

MIX
Paper from
responsible sources
FSC® C013056

Contents at a Glance

Table of Contents

Introduction

● ●

*W*elcome to the world of cognitive behaviour therapy (CBT) for obsessive-compulsive disorder (OCD). Becoming your own expert on OCD and how to overcome it helps whether you are processing a recent diagnosis, trying out some self-help, working with a therapist, taking medication or perhaps thinking of having another go at recovery. On average, a person waits ten years to get help for OCD, but people really do break free from it. We hope this book helps you do exactly that.

About This Book

We wrote *Managing OCD with CBT For Dummies* as a resource for people who are struggling with OCD in some way. Whether you have OCD yourself or you know someone who does, we want this book to help you understand the problem well and to show you how to help yourself (or someone else) tackle the problem.

Understanding OCD is the first step in creating change; know your enemy, and you can be well armed to fight against it. When you know the basics, you can create a clear picture of your own vicious cycle of obsessions and compulsions. In this book, we talk you through how to respond to your OCD differently and offer plenty of tips and practical advice to help you work toward your goal of overcoming OCD. But knowing the theory is unlikely to really change anything. You actually have to commit to (and stick with) experimenting with behaving differently if you want to see change. It's like learning a language; you can study all the grammar rules and even the vocabulary, but you're not going to become fluent unless you start using the language to communicate.

This book is not only about getting rid of OCD but also about recognising what you may be missing as a result of your OCD. It encourages you to look at the bigger picture and think about your life beyond OCD; it suggests you put more emphasis on enjoyable and rewarding activities so you can create the life you want to lead.

Between us, we have over 30 years of experience in helping people overcome OCD, and we have tried to stick closely to evidence-based practice. We're not medical doctors; we've based our advice regarding medication on the United Kingdom's National Institute for Health and Care Excellence (NICE) guidance for OCD (www.nice.org.uk/guidance/cg31). It's compiled by the leading experts in the field, who draw on high quality research to guide their recommendations.

Although this book is meant to help you help yourself, no one expects you to beat OCD entirely on your own. See what support you can get from your doctor and your loved ones and take a look at some of the excellent charities that exist for people with OCD.

Foolish Assumptions

A word of warning: This book is not going to be a perfect fit for your OCD. If you have OCD, our experience tells us that your hope for relief of responsibility and difficulty tolerating uncertainty will mean you will be inclined to focus upon the way in which your OCD 'is different'.

All we can say is that the principles outlined here almost certainly relate to your OCD so focus on applying these to your personal experience of OCD rather than focusing upon your doubts about whether or not this book is 'right' for you.

Icons Used in This Book

To help you pinpoint vital information, we've placed icons throughout the text to highlight nuggets of knowledge.

This icon suggests a practical thing to do or try in order to help you put whatever we're talking about into practice.

The Remember icon prompts you to pause and take note of something that is particularly important and worth committing to memory.

A Warning highlights typical pitfalls that are worth watching out for. Don't worry; it's very common to fall foul of these. That's how we know to point them out to you.

The OCD Demon icon gives an example of the sort of argument your OCD may come up with to try and put you off or derail you. There are many ways OCD tries to do this, so don't worry if yours comes up with a different argument. Just notice that it's your OCD talking and take the words with a bucket of salt!

Tips are practical ideas we hope make your journey of recovery a little smoother.

Beyond the Book

You can pick up extra tips and tools online. Check out the Cheat Sheet at `www.dummies.com/cheatsheet/managin gocdwithcbt` and some short bonus articles at `www.dummies.com/extras/managingocdwithcbt`.

Where to Go from Here

This is a reference book, so you can read it from cover to cover to improve your general understanding of CBT for OCD, or you can just dive straight into the page that interests you most. (This is especially good to remember if you have a strong tendency toward being overly thorough.) The important thing is to use the book in the way you find most helpful.

The best place to start is Chapter 1, which introduces CBT for OCD. If you already understand the basics, you may want to jump straight to Part III for details on combating OCD. Lots of topics in CBT for OCD are interrelated, so we provide cross-references throughout the book to point you to other relevant chapters. You can go straight there or save the related information for later – whatever works best for you.

Part I
Understanding OCD

getting started
with

managing
OCD

For Dummies can help you get started with lots of subjects.
Visit www.dummies.com to learn more and do more with
For Dummies.

In this part. . .

✔ Become familiar with the nature of obsessive-compulsive disorder (OCD).

✔ Recognise how cognitive behaviour therapy (CBT) can help treat your OCD.

Chapter 1

All about OCD

● ●

In This Chapter

▶ Explaining obsessive-compulsive disorder (OCD)

▶ Exploring whether you may have OCD

▶ Getting an overview of what's keeping your OCD going

● ●

*O*bsessive-compulsive disorder, usually referred to as OCD for short, is a common disorder that affects many people all over the world. For a long time OCD was considered rare, but research shows that between 2 and 3 percent of people will likely suffer from OCD at some point in their life-times, so you aren't alone!

In this chapter, we explain what OCD is in more detail and help you ascertain whether the problems you're experiencing are likely to be OCD.

Knowing What OCD Is and Isn't

Obsessive-compulsive disorder (OCD) is characterised by obsessions (see 'Observing Obsessions' later in the chapter) and/or compulsions (see 'Clarifying Compulsions' later in the chapter) – most commonly both. Someone with OCD who doesn't experience both obsessions and compulsions is actu-ally quite rare; however, sometimes people are aware only of their compulsions, such as washing or checking, and no longer notice the obsessions that drive these. Similarly, some people may be aware only of experiencing obsessions and not realise that they're performing internal, mental compulsions. (For more on mental compulsions see, Chapter 5).

OCD ranges in severity from causing distress and negatively impacting your everyday routine to being totally debilitating to the point where you're unable to function normally.

Contrary to popular belief, OCD isn't simply a disorder where people wash their hands too much, check things or keep things orderly. You may have heard people say, 'I'm a bit OCD', usually referring to a tendency for liking things clean or tidy; however, people can have a strong preference for things to be in order and not have OCD. In these cases, people find their preference for cleanliness or orderliness a helpful attribute from which they may often derive satisfaction.

People without OCD commonly have moments of doubt – 'Did I turn my hair-straightener off?' – that lead them to double-check. This tendency is part of being human and doesn't mean you have OCD. If, on the other hand, you repeatedly check the item in an attempt to feel absolutely certain, then you may well have OCD.

OCD is a complex and often debilitating disorder the sufferer doesn't find useful or enjoyable. People suffering from OCD tend to feel high levels of discomfort, often in the form of anxiety, guilt or disgust. People with OCD often have an overinflated sense of responsibility for preventing harm and tend to feel high levels of doubt and uncertainty. A person with OCD tends to know that his behaviours or responses to his obsessions are ridiculous but feels powerless to stop performing them.

Some problems are considered part of the OCD family but aren't strictly the same thing as OCD:

- ✔ **Body dysmorphic disorder (BDD):** A distressing preoccupation with the idea that one is ugly

- ✔ **Hoarding disorder:** A life-interfering compulsion to hoard objects

- ✔ **Hair-pulling disorder (trichotillomania):** A compulsion or impulse to pluck hairs from the body

- ✔ **Skin-picking disorder:** A compulsion or impulse to squeeze spots or pick areas of the skin

- ✔ **Health anxiety:** A distressing preoccupation with the idea that one is ill (usually despite being given medical reassurance) or fear that one will become ill

If you think one of these conditions better describes your problem, we suggest seeking additional advice that is specific to that problem.

Deliberating upon the Diagnosis

The following is a screening questionnaire from the International Council on OCD and can give you an indication as to whether you suffer from the disorder:

- ✔ Do you wash or clean a lot?
- ✔ Do you check things a lot?
- ✔ Is there any thought that keeps bothering you that you want to get rid of but can't?
- ✔ Do your activities take a long time to finish?
- ✔ Are you concerned with orderliness or symmetry?

If you answered yes to one or more of these questions *and* it causes significant distress *and/or* it interferes in your ability to work, study or maintain your social or family life or relationships, then there is a significant chance that you have OCD. For a diagnosis, discuss your symptoms with your doctor.

Considering Causes

The question of how somebody ends up developing OCD has no simple, precise answer. OCD is a combination of several factors: biological, personality, environment and life events. Like so many other kinds of psychological problems, no single type of person develops OCD. We've met people from all walks of life who have OCD, and it certainly has nothing to do with being weak or crazy. However, researchers have identified that some psychological traits tend to be associated with vulnerability to OCD:

- ✔ Perfectionism
- ✔ Tending to be overly responsible
- ✔ Overestimating the importance of thoughts
- ✔ Intolerance of uncertainty

For most people, OCD is probably best understood as a misunderstanding of how their minds work, which can lead to some attempts to solve the problem that backfire. Throughout this book, we help you see how you too may have been trying to solve doubts, intrusive thoughts and uncomfortable feelings and that your solutions may very well have become the problem.

Observing Obsessions

Obsessions are defined as unwanted, recurrent, intrusive thoughts, impulses or images that are associated with marked distress. They aren't simply excessive worries about real-life problems and tend to be the opposite of the kinds of thoughts the individual wants to have. A person with OCD tries to avoid triggering, ignore, suppress or neutralise (for example, try to cancel out) his intrusive thoughts.

The following are common examples of obsessions in OCD:

- Doubts about causing/failing to prevent harm related to dirt, chemicals or germs
- Fears of causing harm to elderly/vulnerable people
- Fear of imagining or wishing harm upon someone close to oneself
- Impulses to violently attack, hit, harm or kill a person, small child or animal
- A need to have certain items or possessions symmetrical or just so.
- Blasphemous or 'inappropriate' religious thoughts
- Fear, guilt or disgust at inappropriate sexual thoughts

The most likely thing your certainty-demanding OCD demon will say to any list of obsessions is 'It's not quite the same as mine; what if something more dangerous than OCD is going on with me?'

Clarifying Compulsions

Compulsions are repetitive behaviours or mental acts in response to obsessions, aimed at reducing distress or doubt or preventing harm. Common compulsions (often referred to as *rituals*) include things like washing, checking, ordering, seeking reassurance, tapping, repeating phrases or actions, saying prayers, replacing bad thoughts or images with good ones and trying to control thoughts. Over time, compulsions become less effective, and people find that they need to work even harder to get a similar result.

✔ The more you check something, the more responsible you feel.

✔ The more you check something, the more doubts you have.

✔ The more you seek reassurance, the less confident you are in your own judgment.

✔ The more you seek reassurance, the less tolerant you are of uncertainty.

✔ The more you suppress a thought or image, the more intrusive it becomes. This works in exactly the same way as trying to get an annoying song out of your head; it works backward and makes it more intrusive.

✔ The more you analyse a thought or threat, the more significant your brain thinks it is and the more attention your brain pays to it. This tendency can mean that thought or threat really dominates your sense of what is happening in the world.

✔ The more you try to reduce and avoid threats (such as contaminants or knives), the more aware of them you become.

Acknowledging Avoidance

A third key aspect of OCD is avoidance of the triggers for obsessions. Each time the person avoids a situation or activity, the behaviour is reinforced because he has prevented himself from experiencing anxiety and the harm he thinks could have occurred. For example, if you avoid touching an item because of the fear of danger or harm, you prevent yourself from feeling anxious, and your mind is likely to encourage you to avoid touching it again.

The more you avoid something you're afraid of, the more your fear of that person, place, object, thought, image, substance or bodily sensation increases.

Avoidance often seesaws with compulsions; if you can't avoid, you'll often carry out a compulsion. If doing compulsions becomes very troublesome, you'll probably try harder to avoid.

The places, people, items, substances, pictures, news stories and so on that activate your obsessions or that you tend to avoid are commonly referred to as *triggers*. Spending a few days with a notebook (or your smartphone!) recording your triggers, what obsessions they activate and how you respond to help yourself feel better (which is likely a compulsion) is well worth the effort. This task is the first step in more clearly defining your problem, which in turn leads to far better solving of that problem.

Calculating the Chances of a Cure

Understandably, a question that many people ask when they get a diagnosis of OCD is 'Can it be cured?' As with so much in recovering from OCD, don't take a black-or-white view on curability. Our experience is that many people who have asked this question have been told 'No. OCD can be improved with treatment, but it's a problem that you have to learn to live with'. We would encourage any person with OCD to take this answer with a very large pinch of salt.

The key is to really understand what a full recovery from OCD means:

- ✔ Your intrusive thoughts have become far less frequent, last for far less time and are significantly less intrusive.

- ✔ You rarely experience clinically significant levels of distress related to obsessions.

✔ Your fears/obsessions no longer prevent you from engaging in important areas of your life.

✔ Obsessions no longer occupy your mind for more than an hour a day.

✔ You rarely find yourself engaged in compulsions such as checking, reassurance seeking or decontaminating.

✔ You no longer avoid triggers because of your obsessions.

✔ You typically normalise and are willing to experience your thoughts, images, doubts and urges. However, you tend not to focus upon them.

✔ Your life has become far more guided by your true hopes, dreams and values as a person rather than by your need to avoid a catastrophe or worry.

The bottom line is that if you follow the principles we outline in this book, work hard to regain your mental fitness and flexibility and refuse to participate with your OCD, the chances of breaking free from the oppression of OCD are excellent.

Chapter 2

Introducing CBT for OCD

In This Chapter

▶ Defining cognitive behaviour therapy (CBT)

▶ Looking at the CBT model for OCD

▶ Beginning to see how your own OCD is maintained

▶ Getting ready to stand up to your OCD

*T*he research-proven treatments for OCD are *cognitive behaviour therapy* (CBT), which includes *exposure and response prevention* (ERP), and high-dose antidepressants. The evidence suggests that the best long-term results come from the correct, OCD-specific type of behaviour therapy or cognitive behaviour therapy. This usually involves some form of deliberately facing your fears and stopping your rituals. You can do it with or without medication, depending on what you and your doctor decide.

This chapter introduces some of the key psychological principles that you can use to fight your OCD and win.

Becoming Familiar with CBT

CBT has become the evidence-based treatment of choice for a huge range of problems. It's a practical, problem-solving type of psychological therapy. Like any good problem solving, the first step in CBT is to define your problem well. Books like this one can really help you understand your problem better and therefore devise better solutions. CBT helps people understand their problems, and create solutions, by looking at the interaction between their thoughts (cognitions), emotions, behaviours and physiology.

The 'hot cross bun' model, shown in Figure 2-1, is the classic way of illustrating how your thoughts, feelings, behaviours and physiology interact.

The following sections break down these pieces.

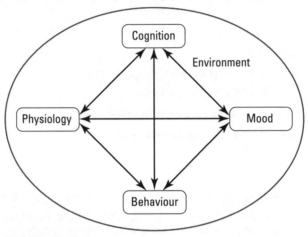

© John Wiley & Sons, Inc.

Figure 2-1: The hot cross bun model.

Cornering cognition

Cognition includes thoughts, images, beliefs, attitudes, attention and memory. An example of cognitions to target for change are misinterpretations of intrusive thoughts (for example, feeling that having a bad thought means you're bad and dangerous or that having a thought about harm coming to someone means you're responsible for preventing that harm). A core concept of this book is that OCD is the product of normal doubts, urges and intrusive thoughts/images that people misinterpret or misprocess in some way. Excessive responsibility, intolerance of uncertainty, placing too much importance on thoughts or images and lack of trust/confidence in your own judgment or memory are all key dimensions of cognition in OCD.

Examining emotion

Examples of emotions that are often prevalent in OCD are anxiety (such as that triggered by fear of being responsible for causing harm or failing to prevent harm), guilt (such as that triggered by the idea that you have not been careful enough or that you're bad, dangerous or have sinned) and disgust (such as that triggered by encountering a bodily fluid or having a morally repugnant thought).

Pinning down physiology

An example of a physiological response in OCD is having sweaty palms when you're anxious (which can be why people with contamination OCD report that their hands feel contaminated when their OCD is triggered). Some people may misinterpret or monitor bodily sensations that are consistent with their fears. For example, someone with a fear of inappropriate sexual thoughts or images may monitor sensations in their genitals. This scrutiny means that the person is more likely to detect minor physiological changes and misinterpret this as arousal, thereby reinforcing the erroneous belief that they are bad or abnormal.

Your doctor may offer you a physiological treatment (usually a high-dose antidepressant) to help with your OCD symptoms. It's important to understand that this does not mean that there is something inherently wrong with your brain, just that such medications have been demonstrated by research to help with OCD.

Buttoning down behaviours

The range of behaviour that people use to cope with their obsessions and fears is absolutely vast. To know whether an action is a problem, the key is to understand the function of the behaviour. If it is aimed at reducing uncertainty removing an unwanted thought/image or avoiding discomfort such as guilt or anxiety, it is very likely to be part of your OCD maintenance.

Observable behaviours such as avoidance and compulsions are often the most noticeable and directly accessible elements of OCD. Examples of these observable behaviours include the following:

- ✔ Checking the gas cooker or boiler
- ✔ Checking whether doors or windows are locked
- ✔ Checking electrical items
- ✔ Counting out loud
- ✔ Repeating actions
- ✔ Saying or repeating a phrase
- ✔ Putting things in order or 'just so'
- ✔ Seeking reassurance
- ✔ Washing/cleaning your hands, body and/or clothing
- ✔ Wiping/cleaning objects, surfaces, floors or door handles
- ✔ Trying to avoid touching or coming into contact with certain substances, objects, places or people

However, some of the behaviours that you're using (and that are maintaining your OCD) are unseen. These are referred to as covert or mental compulsions such as the following:

- ✔ Mentally checking or reviewing actions
- ✔ Counting in your head
- ✔ Repeating phrases
- ✔ Trying hard to make yourself remember something that you checked by staring or saying a phrase in your mind
- ✔ Trying hard to make sure that you have understood what someone has said

⊮ Changing thoughts or images to make them feel safer

⊮ Investigating or elaborating on those thoughts or images

⊮ Trying to push a thought or image out of your mind

You can find a more in-depth explanation of mental compulsions in Chapter 5.

Remember, your OCD demon is constantly looking to highlight that your obsessions are different in some way and that any list doesn't apply to you. Try to see through to the common themes and similarities.

Inspecting interactions

As the hot cross in Figure 2-1 illustrates, the cognitive, behavioural, emotional and physiological elements of human psychology all overlap and interact with each other. For example, understanding the effect of your emotions on your OCD can be helpful. How do your emotions drive the kinds of thoughts that your brain produces? How does your emotional state change how plausible and realistic certain thoughts seem? Consider the effect that watching a really scary movie has on your thoughts about going into the kitchen to make a cup of tea. Because your anxiety is higher, odds are you'll be far more likely to imagine that an axe murderer is in the kitchen and to find dismissing the idea much harder.

Here are a few examples of ways in which your thoughts, emotions, behaviours and physical feelings can interact:

⊮ Anxiety may increase the vividness, plausibility and/or frequency of catastrophic thoughts and images.

⊮ Guilt may make you more likely to think of yourself as bad and imagine that you will be punished.

✔ Checking overloads your short-term memory and makes it harder to feel sure.

✔ Focusing on areas of your skin that you fear have become contaminated may make that skin feel different.

✔ Reassurance seeking, or transferring responsibility to someone else, decreases your trust and confidence in your own mind.

✔ Avoidance and escape (usually compulsions in OCD) maintain fear.

As you can see, the key to understanding and defeating your OCD is to realise that the solution is the problem.

The opposite side of the same coin is that the more you think, assume and act as if you're free from anxiety, the more your brain is likely to adjust its emotional and physiological responses accordingly.

Imagine you have an *OCD-free twin*, who is the same as you in every way but free from excessive fears, compulsions and avoidance. This can be a great reference point when thinking how you could change your behaviour and resist participating with your OCD – ask yourself the question, 'what would my OCD-free twin do in this situation?'

The following is a table of common types of OCD, with illustrative examples of thoughts, beliefs, avoidance behaviours, compulsions and unintended consequences (the way they feed back to the problem). Please remember, these are just examples to give you some ideas of the kinds of experiences other people with OCD have and a few illustrations of how avoidance and compulsions keep the problem going and may in fact make other aspects of your life worse. To really tackle your own OCD head-on, use these to get the 'the solution is the problem' principle clear in your mind and really think through what may be keeping your OCD going.

Example of Obsession	Common Beliefs and Misinterpretations	Examples of Unhelpful Avoidance	Common Compulsions	Examples of Unintended Consequences
Physical contamination (for example, fear of dirt, germs, chemicals or other 'toxic' substances)	Failing to prevent harm is as bad as causing it. I must do everything I can to avoid being contaminated (or contaminating others). If I am to be able to put my mind at rest, I need to be 100% sure that I am not contaminated (or that I have not contaminated others).	Avoiding touching or coming in contact with sources of contaminants (such as toilets, cleaning products or raw meat); employing a safety strategy such as wearing gloves, touching with elbow instead of hand and so on	Cleaning, wiping, laundering, washing hands, showering, discarding contaminated objects.	Financial strain (from purchasing excessive cleaning products or throwing out contaminated objects). Limited activity. Strain on interpersonal relationships. Increased fear. Increased attention/vigilance to potential contaminants in the environment and on yourself, leading you to feel more contaminated.
Thought contamination (for example, the idea that a thought can become attached to or pollute an object or person and cause harm, bad luck or negative changes)	My mind must be clear of any bad thoughts for me to be sure that things will be okay. If I do something with these thoughts in my mind, something bad may happen and it will all be my fault.	Avoiding possible triggers for unwanted thoughts; avoiding taking action until your mind is clear of unwanted thoughts	Repeating actions until done without inappropriate thoughts/images in mind. Neutralising bad thoughts/images by replacing them with appropriate or good ones. Trying to stop thoughts from entering your mind or to erase them from your mind.	Loss of time/productivity. Social embarrassment caused by compulsions. Loss of pleasure. Limited activities. Increased anxiety and attention to unwanted thoughts. Conflict between your apparently superstitious behaviours and your beliefs. Increase in intrusiveness of unwanted thoughts because of thought suppression.
Inappropriate/ unwanted sexual thoughts	Having these thoughts means there is something wrong with me or that I want these things. I need to be completely certain that this is not the case.	Avoiding sex and/or masturbation; trying to avoid becoming sexually aroused.	Looking for material regarding the potential appropriateness/ inappropriateness of thoughts/images. Seeking reassurance that this is 'normal'. Comparing yourself to sex offenders. Monitoring yourself for appropriate or inappropriate arousal. Monitoring what you pay attention to. Trying to stop thoughts entering your mind or to get rid of them in some way after they're there.	Limited activities. Loss of pleasure. Increased doubt. Increased attention to unwanted thoughts or subjects. Thought suppression increases intrusiveness of unwanted thoughts.
Relationship	I need to be completely certain that I am with the right person.	Avoiding activities, physical closeness or separation; avoiding looking at attractive others.	Monitoring romantic feelings toward partner. Comparing those feelings with feelings toward others to scrutinise partner's flaws to work out whether partner is right for you.	Loss of pleasure. Increased doubt. Increased tendency to notice negative things about relationship. Increased preoccupation with fears regarding relationship interferes with following true values about being a good partner.

(continued)

(continued)

Example of Obsession	Common Beliefs and Misinterpretations	Examples of Unhelpful Avoidance	Common Compulsions	Examples of Unintended Consequences
Fear of violence	I may be a risk and I need to do everything I can to minimise this chance and be sure that I won't cause harm.	Avoiding objects that could be used as weapons such as knives, scissors and garden tools; avoiding being alone with people you regard as being vulnerable.	Monitoring yourself for violent thoughts and urges. Rationalising thoughts and trying to convince yourself you're good or harmless. Seeking reassurance from others. Checking for profiles of violent or dangerous personalities. Keeping your hands in your pockets. Trying to keep tight control of temper and behaviour.	Limited activities. Low mood. Increased doubt. Increased sense of being a potential risk; reduced wider view of self. Risk of becoming distant from loved ones.
Fear of accidental harm	If I have the thought that I may have caused harm, it is irresponsible to dismiss it.	Avoiding driving at night; giving someone a wide berth when walking past them on the stairs or near a road.	Repeatedly checking that you have not caused harm (for example, by retracing your steps or replaying activities in your mind).	Limited activities. Loss of productivity. Increased attention to unwanted thoughts/images. Increased doubt due to trying to prove that something did not happen.
Sensory motor	I need to control my breathing/swallowing/blinking and so on or they will not function properly (and death or disability will occur).	Avoiding tasks/activities that you think would interfere with this process.	Monitoring reflex physical processes and trying to control them or check that they are still working.	Limited activities. Loss of pleasure. Reduced productivity. Low mood. Increased awareness of problem area; detection and misinterpretation of sensations.
Order/ symmetry	The only way I can feel comfortable is if everything is just right; otherwise, it's going to bother me all day.	Avoiding having people over to house; not letting children or others near ordered possessions; not using rooms/items.	Straightening. Putting things in size or colour order. Placing items in certain positions until they feel right	Limited activities. Reduced productivity. Self-perpetuating cycle of putting things in order and feeling that ordering is important. Loss of pleasure/satisfaction. Increased irritation and tension with loved ones who may disturb or not conform to your order/symmetry.
Religious/ blasphemous	If I have a blasphemous thought, I must have meant to think it. I'll be punished. A blasphemous thought is as bad as an immoral act.	Avoiding attending place of worship; avoiding highly religious people; avoiding reading religious text or praying.	Trying to live by religious principles to the letter. Replacing blasphemous thoughts or images with more acceptable ones or trying to push these thoughts or images out of your mind. Atoning or making amends for any perceived sins.	Increased frequency and intrusiveness of thoughts. Loss of ability to pursue religious values and practices. Lower mood.
Superstitious	Thinking of a bad thing means it's more likely to happen. Failing to make effort to ward off bad luck is the same as wishing for bad things to happen. I cannot take the risk that if I ignore a thought, something may go wrong. I couldn't live with the guilt.	Avoiding numbers, symbols, people, places and so on that you associate with bad luck or tempting fate.	Carrying out rituals you believe will ward off bad luck. Suppressing or replacing superstitious thoughts. Repeating activities to avoid carrying activities out an unlucky number of times. Repeatedly touching or tapping something to land on a luckier number.	Frustration and loss of time. Difficulty getting the ritual right. Self-perpetuating cycle of acting on superstitious thoughts and feeling they become more plausible.

Visualising your Vicious Flower

The *vicious flower* is really just a set of vicious circles that help you to define your OCD and clarify how the problem works. It's definitely an exercise for a blank sheet of paper and a pen. Don't be afraid to make it scruffy, have more than one attempt and add to or change it as you begin to understand your OCD better.

1. **Pick a good recent example of your OCD being triggered.**

2. **Note the main catastrophic thought/doubt/image/ urge that was triggered (this is your leaf).**

3. **Identify how you interpreted that mental event. What did it mean to you?**

4. **To start building your vicious flower, first identify the output of that interpretation.**

   For example, the output may include some of the following:

   • Emotions

   • Compulsions

   • Mental activities

   • Avoidance behaviours

   • Changes in where you focused your attention

   • Reassurance seeking

   Each petal starts off like the one shown in Figure 2-2a.

5. **Close the loop (see Figure 2-2b) with a comment on how that output may be feeding back in and contributing to the maintenance of the problem.**

   This is what we refer to as an 'unintended consequence'. If you can't think how it may be feeding the problem, ask yourself how you can find out more. Often you can learn a lot by changing that output in some way and seeing what happens.

6. **For each petal, start to think of ideas on how you could make things different and start to test them out.**

   Your overall vicious flower should now look something like Figure 2-2c.

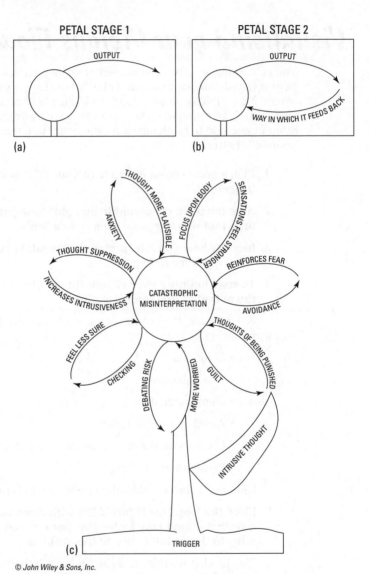

© John Wiley & Sons, Inc.

Figure 2-2: Your vicious flower.

Trying out a New Theory

One of the foundations of CBT for OCD is the idea that there are broadly two theories about what is happening:

> ✔ Theory A: You're at high risk of causing harm or failing to prevent harm. Your solution is to try hard to reduce this risk. (This perspective usually maintains OCD.)
>
> ✔ Theory B: You have an excessive worry about causing harm or failing to prevent harm, and your need to find solutions has become your problem and feeds your worries. (This perspective leads to engaging with the process of solving OCD.)

Your OCD demon is very likely to try to catch you in a trap here, demanding that you work out for sure that you have a worry problem and insisting that you should err on the safe side (trying to prevent the harm). This is unfair; you do have the right to take the risk and break free from your OCD.

So the key in recovery is to test out what happens if you treat your problem 'as if' you believe it's a worry and fear problem. Truly commit yourself to solutions designed to eliminate worry or fear in the long term. This will be absolutely pivotal in your recovery from OCD (and really show your mind that you are serious about real and imagined events being treated very differently).

This experimental approach is the very cornerstone of good CBT. CBT is not a toolbox of techniques to be thrown at a problem. Good CBT matches each aspect of the problem with a solution, and you can then dismantle your OCD until it falls apart. And as the hot cross bun in Figure 2-1 shows, having a positive impact on one element of your mental health is very likely to have a positive knock-on effect on other aspects.

Facing your fears (whether in real or imagined ways), stopping rituals and learning to normalise, detach and allow your thoughts/doubts/images/urges to pass of their own accord are the key ingredients for success. Parts II and III get more into the detail of this process, but there is no time like the present to get started!

Part II
Thinking about Thoughts

Pop-up thought or image		Appraisal or meaning		Obsession (or potential obsession)
	+		=	

 Head to www.dummies.com/extras/managingocdwithcbt for a daily measures sheet to help you track your progress in overcoming OCD.

In this part. . .

✔ Identify how thoughts inform OCD.

✔ Understand the underlying beliefs that help create your OCD.

✔ Figure out how to spot the mental rituals that are aiding the maintenance of your OCD.

Chapter 3

Thinking about Your Thoughts

In This Chapter

▶ Differentiating between automatic thoughts and response thoughts

▶ Understanding how a thought can become an obsession

▶ Taking charge of how you interpret your automatic thoughts

*A*lthough OCD is usually associated with external behaviours like washing and checking, these actions are more often than not responses to internal events, namely your thoughts and what you think about your thoughts.

Whatever type of OCD you have, or however your OCD manifests itself, this chapter is relevant to you. OCD is a psychological problem, and thinking therefore is at the root of the problem.

This chapter explores different types of thoughts and thinking behaviours and explains in detail how a thought can become an obsession. Understanding this gives you a solid foundation on which to build your recovery.

Teasing out Thoughts versus Thinking

Distinguishing the different thought processes that you have is key to understanding your OCD better. In dealing with OCD, you need to know there are two distinct kinds of thinking:

automatic thoughts that pop into your head unbidden, often referred to as intrusions, and then the more considered kind of thought where you're evaluating or *thinking about* your intrusive thoughts. These thoughts about your thoughts may feel automatic, but you have much more control over this second thinking process than you may have realised.

Automatic or pop-up thoughts

Automatic thoughts are rather like the name suggests – they're the thoughts, ideas or images that pop into the mind on a regular basis without you deliberately trying to think them. For example, you may be walking along the road, see a purple house and think 'that's a funny colour to paint your house'. You didn't set out to think about this topic, but something in your environment triggered the thought.

People often have thoughts that can't easily be traced to a trigger. For example, you may suddenly think of Auntie Flo while tying your shoelaces; because tying your shoelaces has no relevance to Auntie Flo, you can consider this thought totally random.

Everyone experiences automatic thoughts (usually many times a day) but doesn't notice them most of the time because they're so common and unremarkable that they just pop in one ear and out the other!

Intrusive thoughts

Intrusive thoughts are automatic, pop-up thoughts very similar to the ones described in the preceding section. The difference between intrusive thoughts and automatic thoughts is that intrusive thoughts tend to catch your attention because their content is unusual, weird or alarming, particularly because they usually appear to be out of character. Intrusive thoughts may also be images (for example, of loved ones dying in an accident), feelings (a sense that something isn't right) or urges (to jump off something or yell out). Intrusive thoughts often carry a sense of foreboding, like they're warning you of something or of doubt (coupled with the urge to remove this uncertainty).

The other difference between automatic and intrusive thoughts is that people tend to react differently to intrusive thoughts because of the meaning they attach to them (see the later section 'Response thoughts').

Normalising intrusive thoughts

Pop-up and intrusive thoughts are an everyday occurrence for everyone. It's part of being human. Some people may have more of them or a greater awareness of them than others, but everyone has them to some extent.

Intrusions can be on any subject matter that a person finds alarming or out of character. What's intrusive for one person may be dismissed as an automatic thought by another. Typical subjects for intrusive thoughts include (but aren't limited to) the following: automatic functions (such as blinking), violence, causing harm (accidentally or deliberately), contamination (of self or others), illness and death, scrupulosity, order, sexuality and superstition.

Response thoughts

Response thoughts are the thoughts or thinking that you do in response to your intrusions. The proper name for them is *metacognition*, which simply means thoughts about your thoughts. It's like a second level of thinking. It differs from automatic thoughts in that it's less random and more within your control, at least with a bit of practice. (We cover this process more in Chapter 5.)

Intrusive thoughts versus response thoughts

It's helpful for you to be able to differentiate between intrusive thoughts and response thoughts so you know which ones you can and can't control. The following table shows the difference between intrusive thoughts and response thoughts.

Intrusive thoughts	*Response thoughts*
Pop into head unbidden	Usually come as a response to automatic thoughts
Often appear to pop in randomly	Attach meaning to automatic thoughts
Are uncontrollable	Able to choose not to engage in this process
Can be on any subject	

Here's an example of an automatic thought and response thought in action: Imagine you're busy doing something and a thought pops into your head about calling your mother. This is an automatic thought.

You could respond to this thought by thinking about it, which would then qualify as a response thought. For example, you may think, 'If she's popped into my head, there may be a reason, so I'd better call her and check that she's okay'.

However, you could also respond by not thinking about the automatic thought; to put it another way, you could not respond to it. Maybe you're so engaged in what you're already doing that you barely notice the thought and don't think about it at all.

This second approach happens with lots of automatic thoughts. They're simply the flotsam and jetsam of a busy mind and just pass by or, as the saying goes, they pop in one ear and out the other. Discovering how to allow your intrusive thoughts to pass by is a key tool in helping you overcome your OCD, as we discuss in Chapter 5.

But what if the thought is really important? Your OCD demon is going to try to get you to respond to your intrusive thought, often by trying to persuade you that *on this occasion* the intrusive thought is really important and needs responding to.

Because OCD is characterised by doubt, a need for certainty and often perfectionism, people tend to get tied up in knots about which kind of thought or thinking they're having. Instead of trying to get it right, think of this obstacle as another exercise in figuring out how to deal with doubt differently and challenge yourself to assume you've understood correctly even if you're not absolutely sure.

Following the Path from Thought to Obsession

An intrusive thought, whatever the content or however out of character it may be, isn't an obsession. However weird or out of character the thought may be, this still does not make it an obsession; it is simply an unasked-for automatic event in your mind. It's only when you attach a meaning or importance to a thought (which in turn leads you to react differently) that it becomes an obsession.

Although we refer to a pop-up thought or image here, urges, doubts, feelings or physical sensations become obsessions in the same way.

Introducing the obsession equation

To turn a thought into an obsession, you need more than just a pop-up thought, image, urge or feeling. You need to add an unhelpful negative interpretation.

| Pop-up thought or image | + | Appraisal or meaning | = | Obsession (or potential obsession) |

Whereas

| Pop-up thought or image | + | No appraisal or meaning | = | Passing thought (no obsession) |

When we talk about *no appraisal*, we mean considering the thought as a normal mental occurrence that doesn't merit any further thought, action or attention.

These appraisals are most commonly made up of two parts:

- An idea that you should not have the thoughts
- A belief that having them must mean something bad

Here are some examples to show you what we mean:

Image of harming my baby	+	Appraisal: I shouldn't have thoughts like that. What if it means I want to do it or I might do it? Maybe I'm a dangerous person.	=	Obsession (or potential obsession)
Image of harming my baby	+	No appraisal. Seen as passing thought (even if content is unpleasant).	=	Passing thought, no obsession

An intrusive thought is just a thought that pops into your head randomly and doesn't necessarily mean anything about anything. A thought is just a thought, whatever the content.

Avoiding putting a positive spin on your thoughts

You may wonder why we don't suggest replacing an unhelpful negative appraisal with a helpful positive one. The reason is that the ultimate aim of treatment is to help enable you to barely notice these random thoughts coming in and out of your mind and to cease to care when you do notice them. It's rather like you may not pay much attention to a song playing in your head.

If you practise reacting to the thoughts with a different appraisal, you're still giving them attention by bothering to respond and intrusive thoughts thrive on this attention. This positive appraisal can also become a way of trying to reassure yourself.

So rather than applying the same process but with a different meaning, you need to practise applying a different process:

allowing these thoughts to come and go and take care of themselves without responding. Chapter 5 provides exercises to help you get the hang of this practice.

Learning to let go of using your feelings as a guide

People often use how they feel emotionally or physically as a guide for assessing an intrusive or pop-up thought. For example, if you feel anxious (or have a symptom of anxiety like heart racing or butterflies in your stomach), you take it as a sign that the thought has value and that something must be wrong or some danger impending. If you feel disgust, you assume something needs avoiding at all costs. If you feel guilt, you assume you must have done something bad.

This reaction isn't entirely unreasonable. Strong emotions like anxiety, guilt and disgust were originally designed to help alert humans to a problem so that they could choose what to do about it. Unfortunately, after you've developed a disorder like OCD, this natural self-preservation mechanism starts to over-function and ceases to serve a useful purpose. To reset the balance, you need to train yourself not to respond to these emotions instead of using them to add meaning and importance to your automatic thoughts. For more on the rationale behind this approach and methods for retraining yourself, head to Chapter 6.

Understanding why obsessions centre on out-of-character ideas

When the content of an automatic thought is out of character, people are more likely to notice it as an intrusive thought. They attach an unhelpful negative appraisal instead of just seeing the thought as a common mental occurrence.

Another way to think about it is that OCD likes to get you where it hurts. When people have particularly strong beliefs (for example, about things like cleanliness, order, religion, etc.), their OCD tends to have content related to these beliefs. This connection is likely because the out-of-character thought sticks out immediately (because it goes directly against everything the person holds true) and the person therefore attaches a meaning to it.

After you've attached a meaning to the thought – along the lines of 'I shouldn't have these thoughts; having them must mean something bad' – you're going to be on the lookout for any further thoughts with that content. The more you look for the thoughts, the more likely you are to find them and then engage in unhelpful responses, continuing the cycle.

The easiest way to think about this issue is that if the thoughts weren't out of character, then they'd be unlikely to bother you and therefore unlikely to turn into an obsession. In other words, you're likely to worry most about thoughts farthest away from your beliefs or values as a person.

Chapter 4

Exploring Beliefs and Meanings

In This Chapter

▶ Examining typical beliefs and meanings about thoughts

▶ Dispelling the myth that you can control what pops into your head

▶ Testing how powerful your thoughts are

A common misconception held by individuals with OCD is that they can and should control what pops into their heads. Given that nobody can dictate the kinds of thoughts and images that enter the mind, this idea can lead to an unwinnable war with your own brain. At the core of OCD is a tendency to greatly misinterpret thoughts and images that you experience.

This chapter examines some of the meanings that people typically attach to their intrusive thoughts that turn them into obsessions (see Chapter 3). When you understand that many of the beliefs you've been labouring under are faulty and are making things worse, you can start to notice when you are acting in accordance with these unhelpful beliefs and choose to teach yourself to respond differently.

Uncovering Beliefs that Lead You to Try to Control Thoughts

As we note in Chapter 3, everybody experiences *automatic thoughts* (random thoughts and images popping into their heads) on a regular basis. The primary factor that differentiates this experience from that of an OCD sufferer is that someone with OCD holds negative beliefs about having these thoughts.

Many people grow up with the instruction that they shouldn't say 'bad' things, often coupled with the warning that 'You shouldn't even think things like that!' So perhaps it's no wonder that many struggle not to place importance on the random thoughts that pop into their minds and end up believing they should be able to control whether certain thoughts enter their heads.

In this section, we look at some of the most common unhelpful beliefs about automatic thoughts. We describe different types of *magical thinking* (the idea that your thoughts have power outside your head), and explain how they keep people stuck in the spin-dryer OCD cycle.

I shouldn't have these thoughts (it must mean something bad)

The sense that you shouldn't have the thoughts because you don't like them and they're out of character is a very common belief. Unfortunately, no rules govern the things that can pop into your head automatically (see Chapter 3). Because you can't control whether they come into your head, telling yourself you should not have them makes no sense.

Your OCD is likely to try very hard to reject the normalising of intrusive thoughts: 'Yeah, but if I was a truly good person I'd never have these thoughts; they must mean something'. The fact is they don't mean a thing. You may notice them more because they are against what you want to think, and you may get more because you try to resist them. But they are normal.

We show you how to demonstrate to yourself that it's impossible to continuously control what pops into your head later in the chapter.

I should be able to control these thoughts (I'll go mad)

In a very similar vein to the belief in the preceding section is the idea that you ought to be able to control whether these thoughts come into your head. Although you may occasionally manage to control what pops into your head on a short-term basis, doing so on a long-term basis is impossible. This belief is usually accompanied by the idea that you must be going mad (or will be soon) because you're not able to control your mind.

If I think it, it's as bad as doing it

This is what OCD experts call *moral thought–action fusion* (TAF). It means believing that having the thought is as bad as performing the action. For example, if you have a thought about harming a child, you're just as guilty as someone who does harm a child. Another example of moral thought–action fusion is the idea that thinking about an action means you want to do it and that thinking is as bad as doing.

When you understand that these automatic thoughts are unasked-for mental events that are out of your control, you can see that having them has no bearing on your personal morality and thus be better able to let these thoughts pass.

If I think it, I'm more likely to do it

This belief is the idea that simply having the thought has the power to make you perform an undesired action and makes you more likely to do so. It's what the textbooks refer to as *likelihood thought–action fusion*. For example, if you have a thought of harming someone when you're near a knife, you may believe that this thought will make you pick up the knife and stab someone.

This belief assumes that your thoughts alone determine your actions and doesn't take into account other important factors like desire and motivation.

Think about how hard it is to get yourself to do something when you really don't want to. For example, imagine you're meant to be going for a run but it's cold and raining and you don't like running even at the best of times. Simply having the thought 'go for a run' is unlikely to make you do it. You probably counter this thought with all sorts of arguments and excuses as to why it's not a good time for a run and end up ignoring or rejecting the thought entirely.

Imagine how simple life would be if having a thought could actually make you do the thing you think of. No more need for working yourself up to things, no more procrastination. If a thought of doing your homework pops into your head, you immediately stop what you're doing, walk to your desk and start studying. 'Pay the gas bill' pops into your head, and you turn the car around, drive home, find the bill and pay it. If this concept were true, people's actions would constantly be driven by whatever popped into their heads, and the likelihood of our having managed to create a civilised society would be far lower.

Thoughts don't make you do things; you choose whether to act on your thoughts. 'Don't go to work today' pops into your head, but it doesn't necessarily mean you don't go. You choose whether to listen and act on the thought or to dismiss it. When *intrusive thoughts* (automatic thoughts that seem out of character) have developed into obsessions, you can easily forget that you still have a choice about whether to respond. When you remember this option, the thoughts can start to seem a little less threatening.

If I think it, it's more likely to happen

This means believing that if you think a thought then it is more likely to happen (*probability thought–event fusion*). To believe this suggests that your thoughts alone have the power to impact events – even events out of your control or sphere of influence.

If you'd like to convince yourself that your thoughts, in and of themselves, are harmless mental occurrences that don't have an influence outside of how you choose to respond to them, try this: Start telling yourself (repeatedly) that you're going to win the lottery jackpot. Keep reminding yourself that you're going to win. Buy a lottery ticket and then keep thinking it's the winning lottery ticket until the draw. Then see whether your thoughts have influenced the lottery results.

If I think about it, I'm responsible for preventing it

This thinking error, based on an overinflated sense of responsibility common in OCD sufferers, suggests that if you think about something, you need to act to ensure that it doesn't happen.

You've probably had moments where you've seen a hazard, like a banana skin on the pavement, and thought to yourself that you ought to move it so no one trips over it. If you were in a position to move it, you'd probably do so, but if you were in too much of a hurry or didn't have a convenient opportunity, you may not stop and do anything about it. For people without an overinflated sense of responsibility, either of these responses would be fine. Regardless of whether they moved the banana skin out of the way, they'd think no more of it.

People with an overinflated sense of responsibility would probably react differently. First of all, moving the banana skin to make sure that the hazard was eliminated would become very important. You'd want to do it regardless of whether it was easy or convenient and despite any costs like being late. After you'd moved the banana skin, you'd start wanting to *make absolutely sure* that you had definitely moved it. This urge would most likely lead you to doubt that you had done it and therefore check that you had. When you felt satisfied that you had moved the hazard and carried on your journey, you may well continue to think back to it to try to convince yourself again that you definitely made it safe and no one could trip. Essentially, you're more likely to keep worrying as you take on too much responsibility for others' welfare.

Recognising You Can't Control What Pops into Your Head

You can't control what pops into your head. This is true for everyone, OCD sufferer or not. That is why we refer to these thoughts, whether intrusive or not, as *automatic* thoughts.

If you don't believe you have no power over what you think about, try this exercise: Close your eyes and try very hard not to think about a pink elephant for one minute. Imagine that you absolutely must not think about a pink elephant in any way, shape or form – no images, no thoughts of 'oh no, I need to not think of a pink elephant' and so on.

What happened? Were you 100 percent successful at not thinking about a pink elephant at all? No ideas of trying not to think about it (which is a way of thinking about it), no blue elephants that you were glad weren't pink and no thoughts that tried to come in that you then pushed away?

Usually, when you try hard not to think about something, you're more likely to think about it, for a few reasons:

- ✔ By saying 'Don't think about it', you're priming yourself to think it.

- ✔ By saying 'I mustn't think about it', you're putting yourself on red alert to notice the thought and are much more likely to notice it.

- ✔ You attach importance to whether you think the thought, which means you're more likely to notice if you're thinking it.

Often, people try not to think about the pink elephant by trying hard to think about something else. Even with this approach, you're in a small way in contact with the idea of a pink elephant because you're attempting to keep the thought out. For more about the effects of trying not to think things, check out Chapter 5.

Debunking the Belief 'I Should be Able to Control My Thoughts'

If you continue to believe you ought to be able to control your thoughts, then it stands to reason that you're going to keep trying to do so. However, you're essentially attempting to achieve the impossible by trying to control your thoughts. Perhaps that gives you an idea as to why the process is so tiring and frustrating!

When you think you should be able to control your thoughts but keep finding that you can't, you're adding fuel to the fire that something is wrong with you (or your mind). This idea tends to make people increase their efforts to control the thoughts, which in turn makes the thoughts more frequent and more alarming. Figure 4-1 shows this very vicious cycle.

Vicious cycle of trying to control thoughts

© *John Wiley & Sons, Inc.*

Figure 4-1: What happens when you try to control your thoughts.

The more you try to control the thoughts, the worse the problem is likely to become! Think about how long you've been trying to control your intrusive thoughts and whether doing so has given you anything other than short-term relief from the problem.

Taking Power Away from Magical Thinking

Often, people believe that thoughts can only influence things in a negative way. A bad thought can make something bad happen, but a good thought (like imagining winning the lottery) can't influence whether something good happens.

Assuming that some thoughts have the power to influence the outside world while others don't just doesn't make sense.

Granted, the way that you respond to your thoughts may impact your behaviour, which in turn could influence external events, but the thoughts themselves don't directly affect anything but your responses to them.

Think of a person you care about deeply. Now say to yourself that this person is going to die today. At this point, you're probably reeling and thinking, 'no way'! That's quite a normal response. No one likes to think about a loved one dying, but this exercise is important to show you that even if you deliberately think of this bad thing happening, it's very unlikely to happen (and if it does happen, it is a coincidence rather than evidence that you can make people die by thinking about them dying). If you challenge yourself to do this activity, you can teach yourself that the thoughts and images that pop into your head don't have an impact on external events if you don't respond to them. We present even more ideas and exercises for getting used to thinking unpleasant things without responding in Chapter 5.

Chapter 5

Mental Responses

● ●

In This Chapter

▶ Identifying different mental behaviours

▶ Understanding how hidden compulsions maintain the OCD cycle

▶ Checking out detached mindfulness

▶ Working with exercises and ideas to help retrain your brain

● ●

*A*ll sorts of things go on in your head, often without your noticing. That can make deliberately changing or improving what's going on in there pretty hard.

This chapter is all about *mental compulsions* – the things people do in their heads in response to obsessions. They're unseen mental actions people perform in an attempt to alleviate discomfort or prevent imagined catastrophes.

If you're one of the many people who suffer from what is commonly referred to as *pure O* (the idea of having obsessions but no compulsions), this chapter is definitely going to be useful for you. If, on the other hand, you think your compulsions are primarily external behaviours, you may be surprised to discover how many of these internal 'mental' behaviours you also engage in.

When we talk about 'thoughts', we are also referring to images, sensations and urges. These work in the same way and tend to bring on similar unhelpful responses.

It's not the thoughts but rather what you do with them that maintains the OCD cycle (see Chapter 2 for a much more detailed explanation). This chapter helps you identify the mental compulsions that you most frequently use because being able to spot them is the first step to changing.

Engaging with Thoughts

Obsessions are normal automatic thoughts you've attached meaning to; the most common meaning is that you shouldn't have these thoughts and that having them means something bad. The obsession equation (see Chapter 3) looks like this:

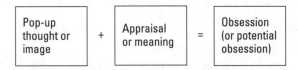

When you believe you shouldn't have certain thoughts, you feel very uncomfortable in their presence and often engage in them as a way of trying to reduce the discomfort.

Unfortunately, the more you engage with these thoughts the worse the problem becomes. Engaging with the thoughts

- ✔ Adds importance to them
- ✔ Feeds the OCD's desire for attention
- ✔ Keeps you thinking about the unwanted subject even more

The following sections highlight some of the most common ways in which people engage with their obsessions, unwittingly maintaining the OCD cycle.

Rationalising

As it sounds, *rationalising* is a way of trying to convince yourself that the obsession is wrong by providing logical arguments against what the OCD is saying. It's similar to a technique used in some types of CBT where you're encouraged to identify 'thinking errors', examining the content of your thoughts and teaching yourself to have a more rational take on them. Unfortunately, this process doesn't help in dealing with OCD. If anything, it makes the problem worse rather than better.

When you try to rationalise your OCD worries, you're engaging with the obsessions and therefore unwittingly feeding them. OCD is like a hungry stray dog; when you feed it, it comes back for more.

However much rational argument you provide, you still won't feel 100 percent convinced. The OCD likes to have the last word, so you could keep arguing with it forever.

If you have a tendency to try to rationalise your fears, notice that the problem hasn't gone away even though you've been doing this for as long as you can remember. That's pretty good evidence that rationalising doesn't help you.

Reassurance

This term refers to when you try to reassure yourself about something instead of seeking reassurance from others (see Chapter 2). This reassurance often takes the form of trying to convince yourself that something isn't true or that something bad isn't going to happen. Rather like rationalising, this is another way of engaging with the OCD that doesn't work in the long term.

 Ask yourself how long you have been trying to reassure yourself. Has this solved the problem? If not, you need to look for a different solution, namely learning to live with the doubt and let it subside on its own.

Checking

Checking is a mental review you perform to try to make yourself feel more certain about something and alleviate your anxiety. Guess what: It works only in the short term. This is why you'll often check more than once. Mental checking functions the same way as overt behavioural checking and is equally unhelpful in helping solve the problem long term. What's particularly important to understand about checking of all kinds is how much information you put into your brain with each check, and how all that checking-driven data overloads your mind. Particularly your short-term memory, which buckles under the pressure and can't move all the information to long-term memory. Combined with intolerance of uncertainty, this is why people with OCD so often seem so unsure about the things they have checked. As far as checking is concerned, less, it would seem, is very definitely more.

Mental checks come in many varieties, including the following:

✔ **Memory checks:** Thinking back over things to be clear that you remember having done something (or not done something), such as locking the door

✔ **Information checks:** Going over information in your head to make sure that you understood something correctly, such as whether the doctor definitely said 'twice a day'

✔ **Emotional checks:** Checking to see whether, say, you feel 'in love' with your partner to convince yourself he or she is the right person for you; common in relationship OCD

✔ **Sensation checks:** Checking to see whether you have any physical sensations in response to something, such as checking to see whether you're aroused

✔ **Thought checks:** Checking to see whether you're still thinking your OCD thoughts

✔ **Response checks:** Analysing your response to a thought or situation to confirm that you still feel an 'appropriate' negative emotion like horror or disgust.

Using mental preparation

People use mental preparation in a couple of ways to respond to obsessions:

✔ **Rather than immediately engaging in an overt compulsion, they employ a hidden compulsion to plan or mentally rehearse a ritual to be done later.** People often use this method in situations where they feel unable to perform the desired ritual, either due to inconvenience or, more commonly, out of shame.

✔ **They plan how to avoid potential triggers, escape from those triggers or at least have a 'witness' present to provide reassurance later if needed.** For example, they plan a route that avoids passing any building works (with a contamination fear) or avoids passing any schools or playgrounds (with a fear of being a paedophile).

Introspection

Introspection is about engaging in the process of trying to understand obsessive thoughts or feelings in order to find reasons for them. Many types of therapy encourage this process, the idea being that understanding the root or cause of your problems helps you get better.

Unfortunately, this process doesn't work in the same way for OCD. When you spend time trying to understand why you have these thoughts or feelings and where they come from, you're simply getting caught in the OCD's trap. Searching for reasons adds more weight to the obsessions, which again maintains the cycle.

Introspection is often used as a way of trying to convince yourself that what you're experiencing is or isn't OCD. For example, you may analyse the content of your thoughts whilst going over old memories to see whether instead of worrying about being gay you're actually gay and you've just been repressing it.

Suppressing Thoughts

The preceding sections cover lots of different ways that people engage with their obsessive thoughts. Trying very hard not to have these thoughts or trying to suppress them in some way is, believe it or not, equally unhelpful.

The main reasons that thought suppression doesn't work are as follows:

- ✔ The more you resist, the more the thoughts persist. It's like putting pressure on a spring; the more you push, the stronger it pushes back.

- ✔ Trying not to think things makes you more likely to think them (see Chapter 4).

- ✔ It reinforces the unhelpful belief that you shouldn't have the thoughts.

- ✔ It helps maintain the erroneous idea that you can't cope with having the thoughts.

- ✔ It gives the thoughts attention and importance they don't merit.

The following sections introduce some of the common things people do to have fewer OCD thoughts, none of which works very well!

Stopping thoughts from entering your mind

When having certain thoughts makes you deeply uncomfortable, doing your best not to experience these thoughts makes sense (on one level). The most common way of trying to prevent certain thoughts is through avoidance, usually of triggers that you think may bring on the unwanted thoughts.

Though avoidance may work in the short term, this behaviour keeps you stuck by reinforcing unhelpful beliefs about the unacceptability of thoughts and your inability to cope with them. (Flip to Chapter 4 for more about this topic.) The more you try to avoid the thoughts, the more you will become convinced that they are bad and that the only reason something terrible hasn't happened is because of your efforts to keep the thoughts at bay.

Pushing thoughts away

As it sounds, this technique is a mental process of trying to force thoughts out of your mind when you notice they're present. Because you don't like the thoughts and think you shouldn't have them, you try mentally to get rid of them by

- Pushing them away.

- Trying to distract yourself with something else so that the thoughts get crowded out of your mind.

- Performing a specific response like clicking on the thought and putting it in an imaginary rubbish bin.

- Saying 'Stop' or pinging an elastic band on your wrist.

- Distracting yourself with something like watching TV. This kind of distraction is different to the helpful alternative of letting the thoughts be there and refocusing your attention elsewhere (see 'Differentiating between Distraction and Redirection' later in the chapter).

Changing or replacing thoughts

Another method people employ in attempting to suppress thoughts that they don't like is trying to change them in some way. Typical examples of this approach include the following:

- ✔ Rewinding and deleting
- ✔ Replacing bad thoughts with good thoughts
- ✔ Neutralising unwanted thoughts with a prayer or magical phrase

This technique reinforces the idea that these thoughts are bad and that having them in your mind is too uncomfortable.

Understanding Why Pushing Thoughts Away Doesn't Work

When you push against something, you're automatically in contact with it, so by trying to push your thoughts away you're achieving the opposite of what you intend.

Go to a door and close it. Push really hard against the door, imagining that keeping the door shut is of dire importance. Keep pushing hard to make sure the door stays shut. Now, whilst pushing the door, try to ignore the door. (This is usually the point at which a quizzical look comes across people's faces.)

Because pushing creates contact, ignoring something you're pushing against is very difficult. A much easier way to ignore the door is to stop pushing it so you're no longer in contact with it. The same goes for pushing thoughts away. When you push, you're keeping yourself more in contact with them and making ignoring them even harder for yourself!

Taking Control of Your Responses

Humans aren't big fans of feeling uncomfortable, so the knee-jerk reaction to discomfort is to try to alleviate it as quickly as possible, rather like scratching an itch. Unfortunately, scratching an itch doesn't stop it from itching for long before you have to scratch it again.

When you recognise that your mental strategy isn't working, it's time to try a different approach. In this section, we introduce you to ways to change your attitude toward your thinking and explain some exercises for getting better at choosing what you pay attention to. Some people find these things alone have a big impact on helping them beat their OCD; however, developing this attitude in conjunction with doing some exercises where you're deliberately exposing yourself to your feared thoughts is often good. For more about how to perform these exercises and specific examples, head to Part III.

Changing your attitude toward your thoughts

Because all mental compulsions or responses maintain the problem, it stands to reason that in order to kick your OCD into touch, you need to respond differently.

Giving up the idea that certain thoughts are 'bad' or 'wrong' and seeing them as internal events (as we discuss in Chapter 3) is one of the keys to overcoming your OCD. If you're willing to test out responding to the thoughts *as if* you believe they don't mean anything, you may start to convince yourself that this is actually the case.

Practising detached mindfulness

The thing most people want to know at this point is how they should respond to or what they should do with their unwanted thoughts instead. The short answer is that you need to learn how to do nothing with these thoughts. You can do so by practising detached mindfulness.

Detached mindfulness or *mindful detachment* is the name given to the process of being aware (mindful) of your thoughts whilst separating yourself from them (detachment). It's about learning to experience your thoughts and feelings and relate to them in a new way. Instead of becoming involved with your thoughts or sensations (or your judgments of these thoughts and sensations), you're taking a step back and becoming a passive observer of these internal events.

Seeing these thoughts simply as internal events helps you let go of engaging in the unhelpful mental responses we discuss earlier in the chapter. Mindful detachment is like being a spectator at a football match (where the players are your internal events) rather than being the referee. You don't need to do anything; you just watch. When you get really good at this you won't even be a spectator, you'll just be like a caretaker at the stadium who is getting on with his job, barely noticing the match taking place.

In practising mindful detachment, it doesn't matter whether the thought, idea or sensation is true or real. Mindful detachment is about relating to these events for what they are (simply internal events) without passing judgment on them. So a thought is a thought whether it's true or not. If someone is standing in front of you and you think, 'There's someone in front of me', the fact that it's true doesn't stop it from being a thought; it's a thought about a fact. You can still see it as a thought and accept it as such without doing anything about it if you so choose. Detached mindfulness is made up of three distinct parts:

- ✔ **Mindfulness:** Developing awareness of intrusions and judgments
- ✔ **Detachment:** Seeing the thought, image, urge or sensation as an internal event that is separate from the self
- ✔ **Disengagement:** Ceasing unhelpful responses like attaching meaning, thought control, avoidance and so on

Remember, if you're willing to try treating your intrusive thoughts as simple internal events that have no particular meaning, then you need to respond to them in the same way as you respond to all your other random automatic thoughts. More times than not, the way you respond to such thoughts is by not responding! (Chapter 3 covers automatic thoughts in more detail.)

Try thinking of your intrusive thoughts as if they're trains passing through a busy station. You may notice them come and go, but you're simply a bystander on the platform and don't need to get on any of the trains. Or think of your automatic thoughts like traffic on a busy road. You don't need to step into the road to try to stop the traffic, nor do you need to wave it past; you just let the traffic do its own thing without getting involved.

We aren't suggesting that you put your thoughts on the trains or cars; that would be the opposite of doing nothing with them! Instead, these analogies are just a way of explaining what we mean by doing nothing with your thoughts. It's not an action; it's an attitude.

Detached mindfulness is *not* about having a blank mind! It's about letting the thoughts be there and doing nothing with them. Detached mindfulness is also not a method of altering or getting rid of your thoughts – it's a way of changing your relationship with them.

Mindfulness meditation

You may have heard the term *mindfulness* in conjunction with the term *meditation. Mindfulness meditation* is currently a popular form of meditation practice aimed at helping people develop awareness of their internal world whilst remaining in the present moment (rather than disappearing off into your internal world).

There are many similarities between mindfulness meditation and detached mindfulness, particularly the idea of learning to watch your thoughts and sensations and let go of responding to them either mentally or behaviourally. Many people find the directive, rigorous daily practice of mindfulness meditation helpful. You usually attend an eight-week course (one session a week) to learn and experience how to practise mindfulness meditation on your own.

When learning any new technique to deal with your OCD, the aim is to help you notice the thoughts in a detached way without engaging in the thoughts in any way. This goal means giving up attaching meaning, ruminating, worrying, trying to suppress or change the thought and so on. Don't use mindfulness meditation as a way of trying to get rid of thoughts or as a coping mechanism to reduce anxiety.

Debunking Concerns You Can't Rule Your Response or Attention

You may well think that your response to thoughts is automatic (not deliberate) and that you can't control what you
pay attention to; these ideas are common. The following sections bust these myths and show you how you can gain power
over these functions.

Regulating automatic responses

Even though your responses seem automatic, you can
respond differently. The first step is being able to identify how
you're currently responding. Familiarise yourself with the
common mental compulsions earlier in the chapter and see
whether you can notice which compulsions you most regularly engage in.

There are different stages at which you can choose to
respond differently:

✔ **Before the compulsion:** This preventative strategy
   involves being aware of your tendency to engage in
   mental compulsions, which allows you to choose not to
   engage in them before they have started. You may still
   feel the urge to perform a compulsion, but you practise
   using this urge as an early warning system that lets you
   choose not to respond to the thoughts. You strengthen
   your ability to do so by deliberately exposing yourself to your unwanted thoughts (turn to Chapters 6
   and 7 for more on deliberately facing your unwanted
   thoughts).

✔ **During the compulsion:** Sometimes you may have
   started to do a mental compulsion before you realise
   it. See this situation as an opportunity to practise inter-
   rupting your mental compulsions rather than completing
   the compulsion to achieve short-term relief. As soon as
   you notice what you're doing, stop, stand back, forgive
   yourself and let the thoughts pass without responding to
   them further.

✔ **After the compulsion:** When your responses seem automatic and you're not yet aware of when you're doing them, you may only be able to notice how you responded afterward. Don't worry; getting sucked into compulsions at the beginning is normal, and even later on, after you've become much better at choosing not to perform compulsions, you can still end up here by mistake. Rather than seeing such a slip as a stick to beat yourself with, use it as an opportunity to reflect on what happened and see whether you can identify the mental compulsions you used. Doing so increases your awareness of how you respond and helps you catch yourself midcompulsion and, in time, precompulsion.

Introducing attention redirection

You probably regularly control what you pay attention to without even realising. For example, think of a time when you've been talking to someone and you've overheard someone else nearby mention your name. At this point, you may have chosen to try to listen to what those other people were saying whilst pretending to continue your original conversation. In doing so, you redirected your attention away from the conversation you were having toward the other conversation in an attempt to hear it. You were choosing what to pay attention to.

In order to do nothing with your thoughts, you want to get better at not paying attention to the fact that they're there. After all, if you didn't notice them, they wouldn't bother you so much. This approach isn't about not thinking certain thoughts; it's about getting better at not listening to them and not thinking about them. It's about caring a whole lot less about them.

Practising Retraining Your Attention

When you have OCD, you've unwittingly trained yourself to be very attentive to the thoughts that you don't like. Luckily, you can retrain yourself to pay a whole lot less attention to these thoughts. Think of it like going to the attention gym.

This exercise uses sounds to help you practise retraining your attention muscles. It strengthens your ability to choose where you direct your attention, making it easier for you to distance yourself from your unwanted thoughts.

To start with, the easiest path may be to set aside 10 minutes a day at home to do this exercise.

1. **Choose four different sounds that you can have competing with each other at the same time.**

   For example, you can turn on the radio or the television, play a song from your phone or computer, put on the vacuum cleaner or run a tap. You can also use external noise like birds singing, traffic passing or building works. Which sounds you choose aren't important so long as you know what you've chosen and the sounds are coming from different sources and different directions. Number the sounds one to four.

2. **Set up the exercise so the sounds are competing with each other; sit with your eyes open and have a soft focus on a point in front of you.**

3. **Pay attention to sound number one.**

   Don't worry if it's hard and you can hear the other sounds; just try to maintain your focus of attention on sound one. The idea is not to block out the competing sounds but to keep redirecting your attention to the chosen sound whenever you notice your attention has wandered off or been drawn away by something.

4. **After about a minute, switch your attention to the next sound.**

5. **Repeat Steps 3 and 4 for the third and fourth sounds.**

You may find that some sounds are easier to tune into than others. That's perfectly normal. Usually, if you find one sound more engaging, like a catchy or familiar song, paying attention to it may be easier than focusing on something less interesting, like the drone of traffic. Keep practising and challenging yourself to get better at maintaining your focus on each of the different sounds.

You may have noticed we didn't give you an instruction not to listen to the other sounds. If you try hard not to listen to something, you end up being more aware of it rather than less aware. That's exactly what happens with your unwanted thoughts. You end up trying so hard not to hear them that you increase your focus on them by pushing them away (see the earlier section 'Understanding Why Pushing Thoughts Away Doesn't Work').

When you've mastered this exercise at home, try doing it when you're out and about. For example, if you're on the bus, you can try focusing on different conversations around you and then switching your attention to the external noise of the traffic. Noise is pretty much everywhere, so you can find plenty of opportunities to practise maintaining and switching your attention.

Having Too Much Internal Focus: The Problem

When you're stuck focusing on what's going on in your head (or your body), you often miss things that are going on around you in the outside world. Perhaps you're busy thinking about something and you accidentally walk into someone on the street because so much of your attention was focused internally that you weren't really looking where you were going or aware of what was going on around you.

This effect can be pretty harmless when it's just the occasional bout of daydreaming, but when you spend every day with the majority of your focus directed inward (on your thoughts, feelings or sensations), it becomes a problem for two main reasons:

✔ Whatever you're focusing on inside yourself takes on more importance and becomes more consuming (and therefore most likely more disturbing).

✔ It prevents you from functioning well in your environment, creating issues with being productive or engaged in other activities (and most likely increasing negative mood).

As you've probably experienced, neither of these things helps you beat your OCD; instead they have the opposite effect and provide a comfortable breeding ground for it.

Redirecting Your Focus Outside Your Head

The first step to figuring out how to focus externally is to start paying attention to your attention. If you don't realise that you're directing your attention internally, choosing to direct it elsewhere is rather hard.

Set an alarm for approximately every hour; when it beeps, just notice what you were paying attention to before it beeped. Was your focus internal or external? Helpful or unhelpful? Do this exercise for a few days or until you develop a greater awareness of where you're focusing your attention. As you get better at noticing without the alarm reminding you, consider making the alerts less frequent – maybe just a couple of times a day. Ultimately, you want to develop your awareness so you're able to notice without needing an alarm to interrupt you from whatever you're focusing on, so think of this trick as a short-term strategy to help you reach this goal rather than an ongoing regime.

The second step is to start practising! In becoming more aware of your focus, you may notice a trend in where you're focusing. For example, you may discover that when you're out doing things with friends, your focus tends to be less internally directed. Or perhaps you notice that when you're sitting alone or on the bus, your focus tends to be more internally directed.

Make a list of the times you're more likely to end up focusing internally and choose these times to practise focusing externally instead. Pay attention to anything outside yourself; that is, ask yourself what is in your environment, whether that's what you can see, smell, hear, touch or feel.

Putting yourself in a role is a way of helping you step outside yourself and pay attention to your surroundings. To help develop and maintain external focus, try one of these (or make up your own):

- ✔ **The forgetful location scout:** Your job is to find locations for a film company that's after a location just like the one you're currently in (be it your living room, a train, the street, your office or whatever). You've found just what the company wants; unfortunately, you didn't bring a camera, so you're going to have to describe the location to the film's director instead. Take in as much detail as possible so you can relay it to the director and give him a really thorough description of the location.

- ✔ **The blind artist's assistant:** You're the assistant of a famous and well-loved artist who has been commissioned to paint the location you're currently in (wherever that may be). Unfortunately, the artist has lost his eyesight, so, as his assistant, your job is to describe your surroundings to him in as much detail as possible so he can paint the environment without ever seeing it. Remember to notice the size, texture and colours of what you can see.

- ✔ **The undercover detective:** You're leading an operation to trap a group of notorious criminals. You suspect that the location you're in is somehow involved. You need to notice as many details as possible – the sounds, the sights, the smells – so you can recall them later in case they're useful as part of the operation.

The more you practise focusing externally, the better you will get at it. Take a few moments to notice how you feel when you spend time focusing outward instead of on your thoughts and feelings.

Another helpful tool for redirecting your attention externally is to set yourself an attention challenge. For example, you may decide that for today you're going to try hard to notice anything around you that is green. It may be the leaves on the trees, the grass, a signpost, a poster or a toothbrush. You can do the same thing with keeping an eye out for a certain number or for how much rubbish is on the streets where you live.

You may well learn that the more you look for certain things, the more likely you are to notice them, like when someone tells you about a new band or you learn a new word and suddenly you notice it everywhere. Think about the negative impact this tendency has when your attention is directed to focus on your OCD worries and things related to them.

The aim of these exercises is to help you start to shift your focus of attention from your internal world to the external world. Don't use them as a way of avoiding thinking certain things or as a way of distracting yourself (see the following section).

Differentiating between Distraction and Redirection

Many people confuse distracting yourself from your OCD thoughts and feelings (which is unhelpful and compounds the problem) with redirecting your attention to external activities.

The line between the two is quite fine. However, here's a simple way to understand the difference: It's all about your intention.

When you try to distract yourself from your OCD thoughts and feelings, you're trying to get away from them or escape from them in some way. You're trying not to think or feel these things. This approach is a kind of avoidance and only makes the problem worse in the long run.

When you try to redirect your attention elsewhere, you're letting the thoughts and feelings be there but choosing not to engage in them.

In order to redirect your attention successfully, you first need to be aware of how you're thinking or feeling and then choose to see this as an internal event that doesn't need responding to. (If that sounds a lot like the detached mindfulness we discuss earlier in the chapter, it's because that's exactly what it is.) To help yourself not respond to it, you redirect your attention to external activities. However, the intention in redirecting your attention is to help yourself disengage from the thoughts or feelings rather than to get rid of them.

Redirecting your attention is easier when you engage in an activity that moves your body or focuses your mind on a new task. For example, getting up and going for a walk or engaging in a sporting activity is more helpful than sitting on the sofa and reading a book (although reading a book is better than engaging in your obsessions). When you engage in something mentally challenging like doing a crossword or knitting a complicated pattern, becoming focused is much easier than it is if you're doing something that doesn't require so much attention to detail, like knitting a simple, repetitive pattern.

Part III
Actively Attacking Your OCD

OCD Type	OCD WAY	ANTI OCD WAY	examples
Contamination (physical)	To avoid possible contaminants or decontaminate through cleaning areas and washing self	To deliberately touch the things you fear and spread this 'contamination' to other things or areas you think are 'safe or clean'	Touch light switches and door handles and then go to kitchen and touch crockery and cutlery in the cupboards Pick up used underwear and put all over current clothes, bag, phone etc. Hug friends to spread contamination further.
Thought Contamination (mental)	Trying to avoid having thoughts or repeating action until done without 'inappropriate' thoughts in mind	To deliberately think unwanted thoughts whilst carrying out tasks or touching objects.	Walk through door whilst holding the unwanted thought in mind Make a recording of yourself repeating the unwanted thoughts and listen to it whilst carrying out trigger activities
Inappropriate/ Unwanted sexual thoughts e.g Fear of homosexual thoughts, fear of inappropriate sexual thoughts about children, fear of staring at breasts or genitals	To avoid triggers, to search for material regarding the appropriateness/ inappropriateness of thoughts/feelings e.g online, monitoring self for arousal at appropriate/ inappropriate times	To deliberately seek out triggers whilst allowing thoughts or feelings and physical sensations to occur without resisting, changing or monitoring	Put self in situations where trigger individuals are present or use photographs, film, tv programmes, magazines etc that contain images of trigger individuals
Relationship OCD	To monitor romantic feelings towards partner, compare with feelings towards others, to scrutinize partners flaws, trying to work out if partner is 'right' for you	To deliberately put yourself in situations which increase your doubts and choose to tolerate the uncertainty.	Acting 'as if' you believe your partner is the 'right' one, despite doubts: e.g buying flowers, saying "I love you", being affectionate. Write down the imaginary scenario that you are stuck in the 'wrong' relationship and you bitterly regret not being with someone else. Read this repeatedly to yourself.
Fear of violence	To avoid being in situations where you could potentially be violent, avoid objects that could be used as weapons, monitor for violent thoughts and urges	To deliberately put yourself in situations in which you fear you could cause harm and have access to potential weapons	Sleep with knife under your pillow Prepare vegetables using sharp knife with the kids in the room Carry a length of rope in your bag

Head to www.dummies.com/extras/managing ocdwithcbt for some clever exposure and response prevention (ERP) exercises you can try.

In this part. . .

✔ Discover how to face your fears and stop your rituals as part of exposure and response prevention (ERP).

✔ Build your own deliberate ERP programme.

✔ Make a plan to stay motivated and tackle your OCD a day at a time.

Chapter 6

Exploring Exposure and Response Prevention

*T*his chapter is all about a specific part of the treatment for OCD called *exposure and response prevention* (ERP). It's the cornerstone of the treatment process, so it's really important that you understand it clearly and know how to use it yourself. ERP is a tried and tested method of helping people overcome OCD; it has been used since the 1950s and has been shown to be extremely effective when it's done correctly! If you've tried it before and it hasn't worked, don't assume it's not for you. This chapter can help you equip yourself with the understanding and expertise to carry out successful ERP so you can defeat your OCD!

Breaking down ERP

Exposure and response prevention (ERP) has almost mythical status in the field of psychotherapy. Early psychoanalysts were afraid that preventing individuals with OCD from carrying out their rituals would lead them to break down and become psychotic. Even within the field of CBT, it's sometimes seen as one of the more brutal and unsophisticated treatments. Many therapists these days prefer the term 'behavioural experiment'. The truth is that ERP doesn't need to be excessively weird or distressing.

Explaining the term ERP

The *exposure* part of ERP means, quite simply, putting yourself in the face of the things that your OCD leads you to fear or avoid for stretches of time to allow your anxiety or discomfort to come down on its own (a process called *habituation*; see the later section 'Clarifying how ERP works' for a more detailed explanation). For example, if you usually avoid touching something for fear of getting contaminated, an exposure would be to touch that thing repeatedly. If you usually avoid certain situations for fear of having certain thoughts, an exposure would be to put yourself in that situation until you became less bothered by it. We talk much more about how to design your own exposures and give you lots of examples in Chapter 7.

The *response prevention* part of ERP means stopping yourself from performing your usual neutralising response – stopping yourself from doing any of the things you usually do to try to keep yourself safe, make yourself feel better or alleviate discomfort in any way. So if your normal response would be to wash your hands after touching something you think is contaminated, the response prevention would be to not wash your hands and instead use them to touch other things even though they *feel* contaminated. If you respond to thoughts that you consider bad or wrong by replacing them with a thought that you consider good or right, the response prevention would be to instead allow the bad thought to stay in your head and take care of itself.

Seeing why response prevention alone is rarely sufficient

People often ask whether they can just do the response prevention part of the treatment. The short answer is no. Studies have shown that response-only treatment isn't nearly as effective as doing exposures with response prevention. You've probably already tried not responding and discovered it's very difficult. If it was easy or effective you wouldn't be reading this book! Think of doing deliberate exposures as a way to bring on the things you struggle with so that you can repeatedly practise responding differently. This approach helps you give up your rituals on a permanent basis.

In OCD treatment, 'exposure' and 'response prevention' aren't two separate entities that you can choose between. They have to go together or you will not get the benefits. Think of them like a toothbrush and toothpaste: One without the other is of limited use to you!

Clarifying how ERP works

In simple terms, ERP works by facing the things you fear instead of avoiding them. By experiencing the discomfort that comes from confronting your fears without performing any neutralising responses or safety-seeking behaviours, you learn that you can cope, however uneasily, with the feared thing; usually, the more you do this, the easier it becomes. This process is called *habituation*. Figure 6-1 shows habituation in diagrammatic form.

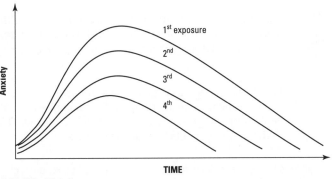

© *John Wiley & Sons, Inc.*

Figure 6-1: Habituation.

Think now about the things you've become used to in the past. For example, perhaps you were nervous on your first day at a new school or a new job. How did you feel by the end of the first week? Did you still feel nervous turning up after six months? Now imagine that because of your nerves, you turned back and went home every time you got to the school gate or the front of the office building. Would you have felt less nervous the following day? Probably not. On the contrary, the more you put off going in, the more your nerves or anxiety about starting would've built up.

Unveiling the secret ingredient: Anti-OCD actions

As we outline in the preceding sections, ERP has two main components, but there is a third part to successful ERP that is often forgotten (perhaps because it's not mentioned in the title). This part is what we refer to as 'anti-OCD actions' or 'doing the opposite of what the OCD wants'.

So to help tackle your OCD using ERP, we want you to

 ✔ Expose yourself to your fears

 ✔ Choose not to perform rituals or neutralising behaviours

Now we add the special ingredient: Deliberately go one step further and do the opposite of what the OCD wants you to do.

 ERP isn't just about not responding; it's about deliberately doing the opposite of what the OCD wants. This *anti-OCD behaviour* is what makes refusing to engage in the rituals easier.

The following sections give you some examples to explain what we mean. These are just a few examples of anti-OCD behaviour. Don't worry if the content of your OCD is different to these; the formula remains the same:

expose to fear + cease rituals/neutralising + anti-OCD behaviour = disempower OCD

See Chapter 7 for more examples of anti-OCD behaviour for different types of OCD.

Paul

Paul has a fear of accidentally causing harm to others. He responds to his OCD thoughts and feelings by checking and rechecking that all the electrical items in his house are turned off and unplugged from the walls. Sometimes it takes him over an hour to leave the house. Paul's anti-OCD behaviour is to deliberately go round the house switching on the lights and leaving the electrical items plugged in before he leaves. This is the opposite of what the OCD wants him to do.

By doing this, he learns the following:

- ✔ To live with the anxiety of having left the electrics on
- ✔ Whether leaving things on leads to his fears coming true
- ✔ That there's no point checking whether things are switched off if he's deliberately left things plugged in and switched on

Gemma

Gemma has intrusive thoughts and images of being burned alive and going to hell. She tries to prevent these thoughts from coming into her head by saying positive prayers and avoids passing anything like graveyards or funeral homes. Gemma's anti-OCD behaviour is to deliberately think of images of hell and the devil as she walks around the graveyard.

By doing this, she teaches herself the following:

- ✔ That she can cope with being in a graveyard *even* if she is thinking about hell
- ✔ That thoughts and images of hell may make her anxious, but she can tolerate the anxiety
- ✔ That she doesn't need to avoid the thoughts and images or things that may lead to them

Sam

Sam worries about contracting HIV by touching something he thinks is contaminated and then passing it on to other people. He avoids many everyday objects such as door handles, light switches, seats and money (to name a few). Whenever he feels contaminated, Sam washes his hands, clothes and body excessively. Sam's anti-OCD behaviour is to touch a feared object and then to not wash and also deliberately spread the contamination by touching his clothes and common objects that he deems 'clean' or 'safe' (such as his phone, remote control, sofa, kitchen table and so on).

By doing so, Sam sees the following:

- ✔ That by contaminating other objects, it's harder to avoid his fear, which gives him more opportunity to confront it instead

  ✔ That his anxiety about the contamination decreases over time even if he doesn't wash

  ✔ That when both he and his surroundings feel contaminated, his desire to wash or avoid actually decreases

Doing deliberate exposures

Deliberate exposure means purposely seeking out the feared situation instead of waiting for a situation to crop up that triggers your OCD. The aim of deliberate exposures is to challenge your OCD head on. Approaching it this way has several benefits. Firstly, it lets you be in charge of your treatment instead of the OCD being in charge of it. You decide how and when to do the exposures. Think of deliberate exposures like trying to get fit for a marathon. You can wait for a situation to arise where you need to run, like being late for a train, to practise your running, or you can join a running club and go regularly.

Secondly, when you decide to do deliberate exposures, you have the opportunity to prepare yourself and prime yourself for how you want to respond differently (that is by not reacting to the thoughts and feelings but instead staying in the situation until your discomfort subsides naturally). If someone snuck up behind you and tried to push you over, she'd probably be quite successful because you wouldn't be expecting it. But if the same person warned you she was going to try to push you over, you'd probably be much better at dealing with it.

Thirdly, when you deliberately do the opposite of what the OCD wants by exposing yourself to your fears and engaging in anti-OCD behaviour, you start to take the power away from the OCD. Imagine that you're standing beside a cold swimming pool and someone (your OCD) is threatening to push you in. You're likely to feel trepidation or fear about the prospect of getting pushed in. However, deciding to take the plunge and jump in yourself gets rid of your anxiety about the prospect of being pushed in; you're probably too busy dealing with the actual situation rather than the threat of the situation. Now that you're in the pool, how powerful is the threat of being pushed in? There's no power in it anymore because the deed is already done.

Think of deliberate exposures as sticking two fingers up to the OCD, as in, 'Really, OCD, you don't want me to do that? Well, tough; I'm going to anyway!'

Fielding Common ERP Questions

Starting something new like ERP can be confusing, and you may have doubts about how to do it correctly. Don't worry; you're not alone! In this section, we answer a number of the questions that we have been asked frequently.

OCD increases the likelihood of doubts and the need for certainty. Be aware of this tendency when you have doubts about the ERP process. Though we're keen for you to be clear about why you're doing ERP and how to do it, there's still an element of trial and error. If you wait until you're absolutely certain you're doing it right, you'll never start, so practise taking a 'try it and see' approach instead.

How should I expect to feel when I do ERP?

Bearing in mind that the idea behind ERP is facing your fears, you should expect to feel at least some level of discomfort or negative emotion, be it anxiety, shame, disgust, guilt, anger or whatever. The idea is to experience these feelings and then choose not to respond. Your level of discomfort may be quite high, but the more you can tolerate it without responding, the better you get.

Sometimes people feel a much lower level of discomfort than they predict. This discrepancy often happens because the fear has built up in their minds to an unrealistic level. If you don't feel much anxiety or discomfort in doing your exposure, you may want to ask yourself whether the exposure is challenging enough for you and look into whether you've properly given up all your avoidance strategies and compulsions. (We discuss these behaviours in Chapter 2). You can read more about troubleshooting in Chapter 7.

Occasionally you may have a more extreme reaction than you predicted. This result isn't uncommon and is nothing to worry about; you can never know for sure how you'll respond, especially if you're exposing yourself to something you've been avoiding for a long time.

How often do I need to do ERP?

As often as possible! Ideally every day, particularly at the beginning of the process so that you can build momentum, iron out any difficulties and start to see some progress. Chapter 8 is dedicated to suggesting a daily plan to help you stick to doing regular ERP. When you've chosen a particular ERP exercise and carried it out, the best plan is to try to repeat the exercise as soon as it's realistically possible. The more frequently you do the exercise, the more quickly you see results.

We understand that you still have your daily life to lead, and you shouldn't do exposures to the detriment of it. However, if you face a fear once and then not again for a while, you can imagine how the fear may build up again. If you're afraid of dogs, patting one dog may help you see you can cope with being near dogs, but if months pass until you see another dog, your fear will probably return. Old habits die hard, so the more rigorously you can challenge them, the better!

Psychological gardening

Imagine that your OCD is a big, overgrown garden full of weeds. If you want the garden to be weed-free, you need to pull up all the weeds. You can pull them gradually a few weeds at a time; however, weeds have a habit of spreading quickly, so the more slowly you pull them up, the more weeds you end up having to get rid of. You may instead choose to try to weed the whole garden in one go, but given the size of the task, you may well end up becoming overwhelmed and giving up or rushing and getting sloppy. The best way to end up with a weed-free garden is to tackle a manageable amount in each session but do the next lot of weeding as soon as possible afterward to avoid too many weeds spreading in between. The more regularly you weed your garden, the more weed-free it becomes. Over time, your garden no longer needs serious amounts of weeding; it just requires maintenance to pull up the odd weed that sneaks back in.

How long will it take for me to get better?

This question is a very hard one to answer because everybody progresses at her own pace. There is no hard and fast rule about how long the treatment should take. It's going to be different for each person depending on where you start, how much time you dedicate to it and how motivated you are to push through the pain barriers. The more effort you put in to helping yourself overcome your OCD (or rather, the more you push yourself to tolerate the discomfort by exposing yourself to your fears and choosing to respond differently), the more quickly you improve.

Some people may find they make dramatic improvement in just a few weeks of dedicating themselves to challenging their OCD; for others, it happens over a course of months or even years. Focus on working through the problem and recognising your progress (as we discuss in Chapter 8) instead of worrying about how long it's going to take.

It takes dedication, perseverance and patience to overcome OCD, and each of these is far more valuable to you than thinking about your speed!

What if I don't have any compulsions?

One type of OCD you may have heard of is often called *pure O*, as in 'purely obsessional'. It may also be referred to as *rumination OCD* because the preoccupation takes the form of ruminating or thinking over and over the same thing. This type of OCD appears to be different to other forms of OCD with obvious compulsions (such as washing or checking), but it's simply that most of the compulsions are mental compulsions (for example, trying to block bad thoughts or replace them with good thoughts). Chapter 5 helps you identify any hidden mental compulsions you may have overlooked. This form of OCD is in fact very common and very treatable, just the same as for other forms of OCD.

After you're clear about what your mental compulsions are, you can use the ERP formula very easily: You cease to do the compulsion and instead do the opposite. For example, if you usually try to replace a negative thought with a positive thought, you not only refuse to replace the negative thought but also deliberately conjure up a thought or image you deem even more negative. (Head to Chapter 7 for more ideas on how to do ERP with mental compulsions.)

What if the problem gets worse?

We understand that changing tactics to deal with your OCD may be a daunting prospect. However, if you follow the steps in this book for dealing with your OCD, it's unlikely to have a negative impact on your OCD. After all, you've been dealing with your OCD your way for as long as you can remember, and it hasn't solved the problem, so it's worth trying to handle it a different way.

Your OCD likes to fight to stay in control, so it may well continue to try to persuade you that this new way is too dangerous or won't work. There's only one way to find out who's right, so try it and see what happens. You may well feel uncomfortable with this kind of uncertainty or risk, which isn't surprising as these are key features in OCD, but tolerating this discomfort and testing out new behaviours is key to overcoming OCD.

Chapter 7

Designing Your Own Exposure and Response Prevention Exercises

In This Chapter

▶ Personalising ERP

▶ Working through the ERP process

▶ Exploring different types of exposures

▶ Overcoming hurdles in your ERP journey

*T*his chapter is the battleground where you train to defeat your OCD. *Exposure and response prevention* (ERP) isn't a hugely complicated process, but the devil can be in the detail, so we guide you through it step by step.

Getting Clear on the Nitty Gritty of ERP

When you know how your OCD is maintained and understand the treatment principles, then you're ready to start tackling your OCD head-on using ERP. Don't put it off any longer; you'll be amazed how much you can help yourself doing this. It can be daunting and difficult, but the good news is it works remarkably well when you do it correctly and stick at it!

Tackling things on your own isn't always easy, but to give yourself the best chance of succeeding, you're going to need a few basic tools to get you started. The main one is a *hierarchy*, a list of your fears that's a little bit like a problems and goals list in standard CBT. You also need a list of your common compulsions. The following sections break down these elements.

Building your own hierarchy

A *hierarchy* is simply a list of your fears, broken down and put into an order of how distressing you find them. Having a hierarchy is a bit like having a map; it allows you to see the lie of the land and work out where you want to go. Tables 7-1 and 7-2 show a couple of sample lists of hierarchy items for you so you can see the idea.

Table 7-1	Example Hierarchy for a Fear of Contamination
Item	*Predicted Distress*
Touching the toilet bowl without washing my hands, touching all 'clean' areas of my home and preparing a meal.	100
Touching the toilet seat without washing my hands, touching all 'clean' areas of my home and preparing a meal.	80
Touching the floor near the toilet without washing my hands, touching all 'clean' areas of my home such as towels, clothes, kitchen surfaces, cups and so on.	60
Touching the door handle and banisters in my home without washing my hands and then touching 'clean' areas of my home such as towels, clothes, kitchen surfaces, cups and so on.	50
Touching a bag that I feel is contaminated and then touching my hair and clothing.	40

Table 7-2	Example Hierarchy for a Fear of Causing Harm by Stabbing	
Item		**Predicted Distress**
Sleeping with a knife under my pillow without telling my partner.		100
Holding a pair of scissors against my partner's neck as part of an agreed exercise.		80
Carrying a pair of scissors with me in my pocket when I'm at home and looking after the children.		70
Carrying a pair of scissors with me in my pocket when I'm at home alone.		60
Handling sharp knives and scissors with the support of my partner.		40

To build your hierarchy, follow these steps:

1. **On a blank sheet of paper, jot down (in no particular order) as many things as possible that you fear doing as a result of your OCD.**

   Don't worry about the length of the list; sometimes people have only a few items, whilst others have hundreds. You're just brainstorming. The number isn't necessarily representative of how bad your OCD is or how long it will take you to get better.

   This list doesn't have to be exhaustive, just a close approximation that covers the main problem areas. You can always add to it as you go along.

   Ask yourself about what you avoid as a result of your OCD. You may also try thinking about what has triggered your OCD over the last days and weeks or ask a friend or family member whether you've missed anything obvious. Look for where you perform compulsions or rituals, and you'll find something that needs to go on your hierarchy.

2. **On a second sheet of paper, write a scale from 10 to 100 on the right-hand side of the page (see the example hierarchy).**

For now, make the numbers on your scale multiples of 10 (10, 20, 30 and so on); you can always squeeze in 15, 25 and other intermediate numbers later if you need to.

3. **Rank your fears from Step 1 on the scale from Step 2.**

First, choose your biggest fear and put it at 100. This fear is the item you wrote down during the brainstorming in Step 1 that you imagine would be the worst possible thing; you're probably shuddering now just thinking of it. If you're just starting out tackling your OCD, you may well write this thing down thinking that you'll never be able to tackle it; this reaction is completely normal. Just stick it down, and you can deal with it later.

Next, choose the fear or fears that are the least anxiety provoking for you and put these beside 10 on the scale. These are the things that you still avoid or respond to unhelpfully but could probably do differently without too much difficulty if you pushed yourself.

Now go through all the other items on the brainstorming list and place them on the hierarchy roughly where you think they come in order of difficulty. This position doesn't have to be exact – it's just a guide – so don't spend too much time worrying about whether something is a 40 or a 50. Ask yourself, 'Do I find this item harder or easier than that one?' By doing so, you get a rough idea of where to put things. You may have several things at the same level of difficulty, and that is also fine.

 The hierarchy isn't a fixed document but rather a guide to help you know what to tackle. As you start doing ERP, you may find things shift around a little in terms of how distressing you find them. You may also find that something simply ceases to be distressing after you've tackled other things of a similar or higher level.

 Items on your hierarchy should be within the realms of possibility. For example, if your worst possible fear is going into space, don't put this at the top of the hierarchy unless you're an astronaut or a space tourist. However, if your fears are based around the vastness of the universe, you could put going on a space simulator at the top instead.

Help! All my fears are equally bad

Occasionally some people find that all their fears feel equally daunting and that it's very difficult to differentiate among them regarding how anxiety provoking they are. If this is the case for you, don't worry; you can definitely still carry out successful ERP (you don't get off that easily)! Instead of doing your hierarchy in order of difficulty, simply put your fears in the order in which you want to start tackling them.

Creating a list of your common compulsions

The second tool for ERP is simply a list of anything that you do to try to alleviate your distress or make yourself feel better when your OCD is activated. These may be rituals, avoidances or compulsions (see Chapter 1 for more on compulsions).

You may think you're pretty clear about all your compulsions, but it's worth having a think about it and writing a list so that you're able to remind yourself in the heat of the moment what you're not allowed to do.

Remember that compulsions can be hidden (covert) as well as obvious (overt; see Table 7-3). For example, replacing one thought with another or with a prayer is a covert (or mental) ritual. For more on mental rituals, flip to Chapter 5. You probably do things that you never realised are actually compulsions.

Table 7-3   Overt and Covert Compulsions

Overt (Visible) Compulsions	Covert (Hidden) Compulsions
Seeking reassurance from others or the Internet	Rationalising
Repeating or redoing actions or phrases	Reassuring self
Excessive washing (for example, of yourself or a family member)	Checking (for example, reviewing details in your head)

(continued)

Table 7-3 *(continued)*

Overt (Visible) Compulsions	Covert (Hidden) Compulsions
Excessive cleaning (for example, of clothes, objects or rooms)	Mental preparation
Avoidance (for example, of people, places or things)	Introspection (for example, repeatedly thinking about reasons for your OCD)
Checking (for example, of light switches, locks, taps and so on)	Pushing thoughts away
Ordering (for example, putting things in place so they feel right)	Trying to stop thoughts entering your mind
Hoarding (for example, keeping papers, unneeded possessions and so on)	Changing or replacing thoughts

We've put together the following sample lists of common compulsions to help you, but when you write your list you need to be specific. So instead of writing 'avoidance', you'd write 'use gloves to empty bins'; you'd write 'keep telling myself I've never hurt anyone' instead of writing 'reassuring myself'. The more specific you are when you list your compulsions, the more easily you can notice when you're doing them or are tempted to do them.

Here's a sample compulsion list for contamination fears:

Use gloves to empty bins

Keep contaminated things in bin liners

Use antibacterial wipes to clean contaminated items

Inspect packaging for signs of contamination

Throw out items I think are too contaminated

Wash hands after touching anything that feels contaminated

Use keys rather than fingers to press lift buttons

Use teeth to pull up sleeves of cardigan

Use antibacterial hand gel when out if I feel contaminated

Search Internet to find out whether something is dangerous and how best to clean things

And here's a sample compulsion list for superstitious fears:

Try to stop thoughts coming into my head (mainly through avoiding things that may bring them on)

Try to mentally block the thought by putting up an imaginary wall in my mind

Exchange bad thoughts with good thoughts

Repeat action until I do it without the bad thought in my mind

Try to reassure myself that I'm a good person and that I haven't done anything to bring myself bad luck.

Ask my mum for reassurance that I haven't acted badly

Repeat my good luck mantra in my head until I no longer feel the effects of the bad thought

The Main Event: Performing Your ERP

When you've compiled your hierarchy and list of common compulsions (as described in the preceding sections), you're ready to dive into ERP. The following sections help you through the process, from choosing a starting point to recognising your progress.

As you begin your exercises, keep these simple rules for successful ERP in mind:

✔ Plan your ERP in advance so you're clear what you're doing and why.

✔ Follow the *challenging but not overwhelming principle* (where you don't push yourself so far that you lose motivation; see the nearby sidebar for more details).

✔ Prepare yourself by reminding yourself of the compulsions you're likely to have.

✔ Be thorough in performing the exercise, remember to go beyond 'normal' and do the opposite of what the OCD wants.

✔ When your distress levels are high, remind yourself that by enduring the discomfort rather than escaping it with a compulsion, you're helping yourself beat the OCD.

✔ Stay in the situation or stick with the exposure until your distress levels lower naturally by themselves.

✔ Be kind to yourself. Notice what you did and congratulate yourself for the effort you made.

We discuss many of these guidelines in more detail in the following sections.

Determining where on the hierarchy to start your ERP

There's no hard and fast rule about exactly where on your hierarchy to start tackling your OCD.

You may feel that the things at the bottom of your list won't stretch you enough and you want to start a little higher up. Or you may feel that even the things at the bottom of your list seem like a big challenge and you want to start there.

We recommend starting with something that is going to be useful to you and make a difference in your daily life. If possible, choose something that you'd feel relieved to be able to tackle differently.

Challenging, not overwhelming

As a general rule, exposures need to be challenging in order to benefit you. It's rather like trying to get physically fit; if the exercise feels really easy, you're probably not pushing yourself enough and therefore not getting much benefit. However, it's best not to tackle something that feels overwhelmingly difficult because doing so is likely to set you back. Imagine you decided to start your new fitness regime by doing a marathon. Most likely, it would be too much for you and you wouldn't be able to complete it. You'd become demotivated, not to mention the potential injuries! (See 'Breaking down tasks that seem impossible'.) In short, you want to feel that the exposure you're doing is a challenge for you but that it's manageable enough not to be totally overwhelming.

Another good way of getting yourself into the swing of doing your ERP is to choose something from the hierarchy that would benefit you in other ways (like going to the hairdresser or spending time with someone your OCD usually has you avoid). When you tackle something that gives you both the benefit of confronting your OCD and a second gain (like enjoyment, getting something done or helping you feel good in some way), the double reward motivates you to tackle other things.

Going beyond what is 'normal'

As we note in Chapter 6, in its simplest form an exposure is doing the opposite of what the OCD wants. This anti-OCD behaviour can often mean doing something that you may think even a 'normal' person (someone without OCD) wouldn't do.

Rest assured that people with OCD are just as normal as those without! We use the word *normal* to refer to someone without OCD only because it's what we often hear OCD sufferers say.

Doing exposures isn't about doing what is 'normal'; it's about taking it one step further so that you can challenge yourself and know that you can still cope even if you go beyond 'normal'. Think of it like having a car with faulty steering. If it naturally steers too much to the right, you can't just put the wheel straight to correct the problem. You have to over-correct to the left. By overcorrecting yourself, you break the habits of feeling comfortable, prove to yourself you can cope and create a buffer so that if you slip back a little, you slip back to the middle ground rather than to the excessive territory where you started.

Breaking down tasks that seem impossible

You may think that some items on your hierarchy are too much to tackle in one go, but that's not a good reason to skip doing it! There's always a way of breaking things down; sometimes it just takes a little bit of thought and creativity. For example, say you have intrusive thoughts that you avoid or neutralise and you find the idea of saying or writing down the thought far too overwhelming. You can start by writing down the first letter or letters of the thought and get used to the feelings that come with doing so before moving on to writing a few more letters. By breaking down the task into manageable exposures, you can work toward writing down or saying the whole intrusive thought.

The idea is to follow the same rules as for any other exposure – that is, expose yourself to a feared stimulus (or a smaller part of it) and then deliberately do what the OCD doesn't want you to do (say it, spread it, touch it or whatever). As long as you don't perform any compulsions or neutralising responses then it is still a worthwhile ERP exercise.

The key to beating OCD is in learning to habituate to the distressing feelings that come with it so that in time the stimuli no longer produce these feelings. See Chapter 6 for more on habituation.

Always bear in mind the 'challenging but not overwhelming' principle (discussed in the nearby sidebar). There's no point in breaking everything down if you don't need to; it only means twice as much work to get through everything.

Working your way through the whole hierarchy

You may be wondering 'Do I have to do everything on my hierarchy?' The short answer is yes. If you expend time and energy challenging many of the things on your hierarchy but don't confront the big things at the top, it's likely the OCD will slowly get worse again. It's rather like weeding a garden but not bothering to get rid of all the weeds. The ones that are left simply pollinate, and the problem returns. Then all the time and effort you put into weeding in the first place was a waste.

As you work through your hierarchy, you may find that day-to-day life becomes a little easier and that your OCD isn't such a big problem. When you're feeling better, it can be hard to motivate yourself to challenge the toughest things. Don't let your OCD persuade you that there's no point in taking on the big stuff!

The only time you don't need to do something on your hierarchy is if you cease to find it distressing. Sometimes people find that when they've started doing ERP on some items on the hierarchy, other items automatically no longer seem to be a problem. If this is the case, then you can simply cross those items off (and give yourself a big pat on the back).

Knowing when you can stop the exercise

Here are a few criteria you may be tempted to use to decide when to end your ERP exercise:

- ✔ When I feel really distressed
- ✔ When I run out of time
- ✔ When I can't be bothered any more
- ✔ Whenever I feel like it
- ✔ When someone else tells me it's enough
- ✔ When the dog needs feeding

Unfortunately, none of these criteria are valid. Realistically, if you're doing your ERP exercise correctly, you're going to be feeling some level of distress. It's human nature to want to try to escape from these feelings (that's what your compulsions and rituals have been for until now).

Bearing in mind that the idea of ERP is to help you get used to these feelings without escaping, it's important to wait until your discomfort levels have dropped as much as possible before stopping the exercise. If your discomfort level starts very high, say a nine out of ten, you may find you can only stay in the situation until it reaches a six or a seven. However, the longer you can encourage yourself to stay in the situation, the more accustomed you become to the stimulus, thereby further decreasing your levels of distress. As your distress drops, your urge to perform a ritual should also decrease.

It may sound strange, but the more thorough you are in doing the opposite of what the OCD wants (for example, spreading the contamination or saying the things you usually avoid thinking), the easier it will be for you not to perform any compulsions. It's as if the OCD thinks 'Whoa, that's way too much for me to put right, so I may as well just lump it and get used to it.'

Exploring Ideas for ERP Exercises

After people understand the theory behind exposure and response prevention, they can sometimes be stuck with getting started on choosing an exposure for a particular fear. However, there is no shortage of ways to do this; you may just have to be bold and creative. There are three main ways of doing exposures:

✓ **Real-life exposure (often referred to as *in vivo*):** As the name suggests, *real-life exposure* is when you're able to confront the actual fears or triggers, whether it's handling something contaminated, going to a gay bar, being near a potentially lethal weapon, walking through a graveyard or whatever.

✓ **Simulated exposure:** *Simulated exposure* is using something to trigger the obsession without actually being in the situation. You use it when putting yourself in the situation isn't possible or in order to give you more access to doing ERP exercises. Examples include watching films, looking at images, going on a flight simulator and so on.

✓ **Imaginal exposure:** *Imaginal exposure* means using your imagination to conjure up the worst-case scenario, which you then write down, read aloud or record and listen back to. This approach is particularly useful when the triggers are specific thoughts or images because it takes the sting out of them. Hearing yourself saying these things helps you get used to them and encourages you to learn to let them be there without paying attention to them.

Your OCD demon is likely to argue that you shouldn't get used to these thoughts because they're too terrible and because not reacting is like saying they're okay to have. Remind yourself that they're just thoughts and that they don't mean anything. For more details on this argument, check out the chapters in Part II.

The following table offers ideas for exposure exercises with different kinds of OCD. Bear in mind that these are just a few examples to help give you an idea. It's by no means a comprehensive list; there are endless ways of doing these exercises. It's worth reading through the ideas for all the different types of OCD (not just your own) because the examples aren't useful for only the subtype listed. They may well be good for you too.

If it feels wrong, it's probably right!

OCD Type	OCD Way	Anti-OCD Way	Examples
Contamination (physical)	To avoid possible contaminants or decontaminate through cleaning areas and washing self	To deliberately touch the things you fear and spread this 'contamination' to other things or areas you think are 'safe or clean'	Touch light switches and door handles and then go to the kitchen and touch crockery and cutlery in the cupboards. Pick up used underwear and put all over current clothes, bag, phone etc. Hug friends to spread contamination further.
Thought contamination (mental)	To try to avoid having thoughts or repeating action until done without 'inappropriate' thoughts in mind	To deliberately think unwanted thoughts whilst carrying out tasks or touching objects.	Walk through door whilst holding the unwanted thought in mind. Make a recording of yourself repeating the unwanted thoughts and listen to it whilst carrying out trigger activities.
Inappropriate/ Unwanted sexual thoughts e.g fear of homosexual thoughts, fear of inappropriate sexual thoughts about children, fear of staring at breasts or genitals	To avoid triggers, to search for material regarding the appropriateness/ inappropriateness of thoughts/feelings e.g online, monitoring self for arousal at appropriate/ inappropriate times	To deliberately seek out triggers whilst allowing thoughts or feelings and physical sensations to occur without resisting, changing or monitoring	Put self in situations where trigger individuals are present or use photographs, film, TV programmes, magazines etc that contain images of trigger individuals.
Relationship OCD	To monitor romantic feelings towards partner, compare with feelings towards others, to scrutinise partner's flaws, trying to work out if partner is 'right' for you	To deliberately put yourself in situations that increase your doubts and choose to tolerate the uncertainty.	Act as if you believe your partner is the 'right' one, despite doubts: e.g buying flowers, saying 'I love you', being affectionate. Write down the imaginary scenario that you are stuck in the 'wrong' relationship and you bitterly regret not being with someone else. Read this repeatedly to yourself.
Fear of violence	To avoid being in situations where you could potentially be violent, avoid objects that could be used as weapons, monitor for violent thoughts and urges	To deliberately put yourself in situations in which you fear you could cause harm and have access to potential weapons	Sleep with knife under your pillow. Prepare vegetables using sharp knife with the kids in the room. Carry a length of rope in your bag.
Fear of accidental harm	To avoid being in situations where you could potentially cause harm. To repeatedly check that you have not caused harm.	To deliberately put yourself in a situation where you fear you could cause harm, reduce caution and eliminate safety behaviours, resist checking	Drive down a busy street without looking in your mirror. Stand shoulder to shoulder or right behind someone on a train platform.
Sensory motor	To monitor reflex physical processes and try and control them or check that they are still working. To avoid doing anything that you think would interfere with this process.	To deliberately do things that you fear would interfere with these reflex processes and let go of monitoring and checking. Do the opposite of what you are trying to do to solve the problem.	If you fear you can't breathe, challenge yourself to hold your breath. If you fear you breathe too much or that your breathing is irregular, deliberately pant or hyperventilate. Write down your worst case scenario and read it to yourself repeatedly.
Symmetry/ordering	To try to make sure that everything is just the way you feel it needs to be. To reduce activities that would interfere with this.	To deliberately make things feel 'wrong' by putting items out of order or out of place	Mess up the order of your books so they are not in size/colour order. Leave the bed unmade.
Religious/blasphemous	To try to live by religious principles to the letter. To replace blasphemous thoughts or images with more 'acceptable' ones or try to push these thoughts or images out of your mind.	To deliberately expose yourself to triggers for blasphemous thoughts	Go to a place of worship and allow blasphemous thoughts to enter your mind or deliberately think blasphemous thoughts. Record yourself saying blasphemous thought or describe blasphemous images and listen to it repeatedly.
Superstitious	To avoid things you think would cause bad luck or disaster. To carry out rituals you believe will ward off bad luck.	To deliberately expose yourself to feared triggers and do things that you think will cause bad luck or disaster.	Break mirror. Open umbrella indoors. Carry out activities in multiples of 13 (or take the no 13 bus).

Dealing with Obstacles

In CBT, we see obstacles as opportunities for learning. There are a number of common obstacles that people tend to encounter when using ERP. In this section, we discuss these pitfalls and give you tips about how to handle them.

Performing a compulsion after an exposure

An exposure is about not only exposing yourself to the feared thing but also not performing the usual response. The easiest way to help yourself not perform your compulsion or ritual is by going that step further and taking an anti-OCD action (see Chapter 6). However, even when you've done everything by the book, your OCD sometimes wins the battle and you go for the short-term fix to lower your distress level.

If this happens, don't worry; you're not expected to necessarily manage it every single time. If you do end up doing a compulsion or ritual, remember three important things that you can do to help yourself in the future.

1. **Notice what happened.**

   Think about why you didn't feel able to resist doing the compulsion. Remember this issue as something that may well come up next time so you can remind yourself not to give in to it.

2. **Repeat the exposure.**

   When you neutralise the exposure by bringing your distress down with a short-term fix (like washing, redoing, reassuring and so on), the best thing to do is to redo the exposure so that the OCD hasn't had the last word! This way, you help yourself become stronger at resisting the compulsions because if you re-expose every time you do a compulsion during an ERP exercise, your mind starts to think there's no point doing the compulsion.

3. **Don't give yourself a hard time.**

   When you beat yourself up, you feel bad; when you feel bad, you feel less able to fight the OCD. Remind yourself that ERP is hard work but that if you keep trying, you'll slowly start winning. Accept yourself in spite of failing at the exposure on this occasion, and you'll be more likely to achieve it the next time.

Working through high discomfort levels

As you know only too well, feeling high levels of distress is very unpleasant, so it's understandable that you want those feelings to lessen as soon as possible. However, on some occasions when you're doing an ERP exercise, you may find that your distress levels take a considerable amount of time to decrease. This time frame isn't necessarily a sign that you've done something wrong or that you should do something to alleviate your distress. Instead, it's a sign that you're challenging something difficult. The more you can stick with these feelings without trying to do anything to make yourself feel better, the more powerful the exercise.

Try these strategies when your distress levels are high:

- ✔ Remind yourself that distress may be super unpleasant but isn't harmful and that this short-term pain helps get you to the long-term goal of overcoming your OCD.

- ✔ Try to focus on an activity rather than on how distressed you're feeling. If you can stay in the situation that you feel distressed about, then so much the better. For example, if you've just contaminated your kitchen, focus on doing an activity there like sweeping the floor, sorting some mail or tidying a cupboard.

- ✔ Make sure you've not only exposed yourself to the fear but also done the anti-OCD actions. It may sound counter-intuitive, but the more you do the opposite of what the OCD wants, the more easily you can habituate to the situation.

- ✔ Note how long you were able to stick with the distress before stopping the exposure or performing a compulsion. When you repeat the exposure, challenge yourself to stay with it a little bit longer than the previous time. If you do this every time, you slowly learn to habituate to the feelings. (You should try to repeat the exposure relatively soon – for example, the next day.)

Troubleshooting when you're not distressed in ERP exercises

In general, people tend to feel distress or discomfort when doing ERP; however, this is not always the case. If you don't feel much distress when doing the exercise, there may be several reasons for this:

- ✔ **You have chosen something which isn't challenging enough.** See the sidebar 'Challenging but not overwhelming' in this chapter for info on this principle.

- ✔ **You're unintentionally performing some kind of safety behaviour (often avoidance) or compulsion, such as reassuring yourself that nothing bad will happen.**

- ✔ **You're doing the exercise correctly, but it's just not as distressing as you expected.** This situation is actually quite common. Often, when you've been avoiding something for a long time, the fear builds up in your mind and you assume it's going to be much more distressing than it is in reality. You can probably recall other situations where you've worried about how you'll cope with something, and then when the time has come it wasn't nearly as difficult as you'd expected.

Anxiety often makes people overestimate how bad something is going to be and underestimate how well they're going to cope with it, so you get a double whammy! In reality, things are often not as bad as you think, and your coping skills are much better than you give yourself credit for. But of course you only discover this by pushing yourself to do things that you fear!

Making time for ERP

Life can be busy at the best of times, but when you're struggling with OCD slowing you down, it can be even harder. Although remembering to fit something new in may seem like a challenge, exposures are rarely more time consuming than the rituals that you're used to doing. 'I didn't have time' is usually another way of saying 'I didn't want to do it', so don't let yourself be fooled by this excuse.

In general, try to get an ERP exercise under your belt as early in the day as you can. By doing so, you put yourself in the 'overcoming OCD' mindset for the rest of the day, and you give yourself the satisfaction of having already done something to tackle it.

On occasions when time is short or there's another obstacle and you can't do the exercise you were hoping to (for example, you were planning on driving somewhere and the car is at the garage), see whether you can find an alternative, briefer exposure to replace it.

Accepting when it doesn't go well

Despite your best efforts, there will probably be times when you struggle to push yourself to engage in your ERP or you're unable to complete the exercise without performing some kind of compulsion or neutralising response. This is completely normal! You're pushing yourself to overcome something difficult and challenging yourself to act in new, uncomfortable ways. You're not expected to triumph every time.

If you're unwell, tired or have lots of other things on your plate, you may find it harder to summon up the strength to battle your OCD. When this happens, the most important thing to remember is to not give yourself a hard time about it. This isn't about letting yourself off the hook and letting the OCD win; it's about being compassionate to yourself and recognising that you're human and sometimes make mistakes or fail. However annoying this humanity is, it's not the end of the world and doesn't mean that you'll never overcome the problem. It just means you haven't managed it on this occasion. Beating yourself up only makes you feel worse and doesn't motivate you to do better next time.

When times are tough, sometimes you need to push yourself harder so you can feel a sense of achievement. At other times, you need to let yourself off the hook temporarily and just do whatever you can to make it through the day. On these occasions, remember to do the following:

✔ Accept that you couldn't manage it on this occasion.

✔ Recognise that setbacks are a normal part of the process and only to be expected.

✔ Notice what was making it extra hard for you.

✔ Think about whether you can do anything to help yourself next time.

✔ Put a plan in place for when you'll redo the exercise.

✔ Be kind to yourself; take some time out to do something for yourself that is relaxing and caring.

Realising the treatment has become part of the problem

Unfortunately, the common issues of needing certainty and perfectionistic tendencies can often transfer themselves onto the treatment process itself. For example, you may check that you're following all the rules to the letter or seek reassurance from a book, website or therapist that you're doing the treatment correctly. This experience isn't uncommon and is a way that your OCD can keep you stuck in the vicious cycle of needing certainty and avoiding the discomfort of doubt. If you notice this has started happening for you, you simply need to follow the same rules to fight back:

✔ Deliberately do the exposure incorrectly and deal with the discomfort without engaging in any neutralising responses.

✔ Remind yourself that responding differently to the uncertainty is an essential part of the recovery process.

The point of ERP isn't to get rid of the thoughts and feelings but to allow you to deal with them differently. If you're no longer responding to thoughts and feelings with compulsions, rituals or neutralising responses of any sort, there is no need to continue the exposures.

Chapter 8

Beating OCD One Day at a Time

*W*ouldn't it be nice if you could change a habit or fix a problem by simply finding out how to do it and then giving it a go? Unfortunately, it's rarely that easy. At the very least, you usually go through a period of trial and error before you perfect a new skill. Imagine if when you learnt to walk you just stood up on two feet one day and were perfectly steady, able to run, hop, skip and jump. As you know, that's not how it happens. First you spend ages watching and understanding and then learn to stand and take a few steps. Slowly but surely, you learn to balance and become more confident as you practise.

The same goes for using cognitive behaviour therapy (CBT) to treat OCD. Yes, it's that old 'practice makes perfect' idea again – and along with practice, you're going to need a bit of good old-fashioned determination and a healthy dollop of patience.

This chapter is all about how to help yourself keep practising your knowledge and honing your OCD-beating skills. We've created a step-by-step plan for you to use and refer to on a daily basis so that you have a framework to help you overcome your OCD.

Making a Step-by-Step Daily Plan

The journey you're undertaking can sometimes be a rough one, and it's easy to get thrown off course, so use the steps in this section as markers to keep you heading in the right direction.

Step 1: Understanding the goal of CBT for OCD

Before you start any new venture, you want to be clear about why you're doing what you're doing, so the first thing you need is a goal to work toward. Otherwise, you can easily get lost along the way.

When we ask people what their goal is regarding their OCD, the most common answer is 'to get rid of it', which of course makes a lot of sense given that it's something that's causing great discomfort and getting in the way of everyday life. However, with that goal you run the risk of falling into the same old traps that OCD sets for you: reacting to intrusive thoughts, images, feelings and urges with the aim of getting rid of them immediately (short-term gain) rather than choosing not to act on them and dealing differently with the feelings that come as a result (long-term gain). (See Chapter 2 for more on the benefits of choosing to react differently to your OCD.) So we propose that you take on a slightly different goal that leads you toward not having OCD without requiring a specific focus on getting rid of it.

As we note throughout the book, CBT works for OCD across the board, so everyone can have a common goal no matter what kind of OCD you have. It may not be as snappy as 'to get rid of OCD', but it's still a good goal for you whether you think yours is the most common or the most unusual OCD:

> To learn to allow all unwanted intrusive thoughts, images, feelings and urges to take care of themselves by training myself to react differently to them until they no longer cause me significant distress or interfere with my life.

The better you get at acting against your OCD, the easier ignoring it is. The easier ignoring it is, the less it bothers you. The less it bothers you, the easier it is to act against it and get on with more important things in your life!

Step 2: Creating milestones

Beating your OCD is a process that takes time, rather like a long journey. Milestones help you

✔ Know that you're heading in the right direction

✔ Feel less overwhelmed by breaking down a big task into more manageable pieces

✔ See how far you've come

One way of creating these milestones is to make a list of the fears you want to tackle and think about where you put your milestones in the list. (Chapter 7 has more on making such a hierarchy.) Perhaps you want to tackle three or four things on your hierarchy to reach the first milestone. You can put in as many milestones as you think are helpful.

Another way of creating milestones is to make a list of things you'd want to be able to do if you didn't have OCD (see Part IV for more on living your life by your values). What do you need to be able to do to feel you've reached a milestone? What you consider a milestone is up to you. Remember that reaching your milestones gives you a sense of achievement and encourages you to keep going, so be generous to yourself and put in enough milestones to keep you motivated. (The later section 'Rewarding yourself' gives you a table to list your milestones so you can keep track of how you're doing.)

Step 3: Planning your ERP

The most powerful treatment method for challenging your OCD is *exposure and response prevention* (ERP). ERP is the process by which you expose yourself to your fears whilst giving up the things you usually do to keep yourself safe or comfortable and refuse to engage in any rituals or compulsions. You can find a more detailed explanation of ERP in Chapter 6.

Committing to act against your OCD is one thing; following through is another. For this reason, we suggest daily ERP practice so you can get into a routine; leave no room for putting it off until tomorrow!

Do yourself a favour and take the easy option of making a plan to practise your ERP so you can stick to it even when you really don't feel like it and your OCD is throwing all sorts of grenades. After all, if you wanted to go and see your favourite band in concert, would you buy the ticket and sort the travel in advance or would you pitch up on the day of to see whether you could get a train to the venue and hope that tickets were still available for the gig? Sure, the second way may work, but it's a lot less likely to.

Think about setting aside a few minutes at the end of each day to plan your exposure for the following day. When you get into the habit of reviewing your daily exposure (see the later section 'Step 5: Keeping track of your ERP'), planning the next one at the same time follows naturally (even if you're going to repeat the same exposure).

As you've unwittingly trained yourself to deal with your OCD in unhelpful ways, it's going to take time and perseverance to retrain yourself. OCD is a cunning beast and will try as many ways as possible to deter you from doing anything it doesn't like – the trick is to make a plan and stick to it no matter what the OCD says or does to try to dissuade you.

Bear in mind the guidelines for successful ERP listed in Chapter 7 when designing your exercises. If you're having any trouble thinking of exposure exercises, you may also want to consult the table in Chapter 7.

Step 4: Doing your ERP

This step seems rather obvious, but if you miss it, you're unlikely to reach your goals.

There are three parts to doing an ERP exercise:

- ✔ Exposing yourself to your fear
- ✔ Resisting the urge to perform any compulsions
- ✔ Acting in an anti-OCD manner

Chapter 7 has a more detailed explanation of how to do ERP exercises correctly.

Your OCD will put up plenty of excuses – no time, too scary, what if . . . and so on – to discourage you from following through, but if you want to change, you need to learn to ignore all these things. (Note that if you call it a 'reason', it's probably still an excuse.)

Step 5: Keeping track of your ERP

Almost as important as doing the exercises (but not quite) is reviewing how they went and what you learnt from the process. The more understanding and awareness you have of what goes on for you when you act against your OCD, the better armed you are to keep challenging it. You're putting yourself through the hard work of training yourself to react differently to your unwanted thoughts, images, feelings and urges, so there's no point skimping on the review that helps you hone your skills and be a stronger opponent to your OCD.

To help facilitate the planning and review process, we've created a table for you. Take a look at the example and remember that you can use this tool regardless of what form your OCD takes. For example, if you don't have any safety behaviours that you know of and only have compulsions (physical or mental), then just fill in the compulsions box and leave the other one blank. Try really hard not to leave the 'what I learned' box blank. If it's the same thing that you learned the day before, just repeat it. If you're struggling to think of something, ask yourself, 'Have I lived to tell the tale?' Knowing that you're getting through it, however uneasily, is a very useful thing to remind yourself.

ERP exercise for today: (what you are deliberately exposing yourself to in order to habituate to anxiety and learn to respond differently)	e.g driving past the local school at collection time when there will be lots of children about		
Safety behaviours to give up: (anything you do to reduce potential anxiety or risk)	e.g driving very very slowly, checking constantly in mirrors	**Did I give up ALL my safety behaviours?** (If not, what did I do and why?)	I drove at the speed limit and only checked in my mirror three times
Compulsions to give up: (anything you do during or after to try and make yourself feel better)	Retracing my steps to check there was no one hurt, checking online that there are no fatalities reported, replaying the journey in my mind to convince myself I didn't hit anyone	**Did I stop myself from doing ALL my unhelpful responses?**	I started to replay the journey in my mind to convince myself I hadn't hurt anyone but once I noticed what I was doing I practised letting the thoughts be there without engaging in them
Prediction 1: **Anxiety level** (how anxious you think you'll feel)	8/10	Outcome 1: Anxiety level	6/10
Prediction 2: **Duration** (how long you think the anxiety will last)	About an hour	Outcome 2: Duration	About 20 minutes
Prediction 3: **Coping ability** (how well you think you will cope)	I'll be a mess; I won't be able to get on with anything afterward	Outcome 3: Coping ability	I felt bad but I managed to go and do the shopping
WHAT CAN I LEARN FROM THIS ERP EXERCISE? It was hard and I didn't want to do it, but it was a bit less anxiety provoking than I thought it would be. The anxiety went down after a short time even though I didn't do anything to try to make myself feel better, and I was able to get on with my day.			
ANYTHING I NEED TO DO DIFFERENTLY NEXT TIME? I need to challenge myself not to look in my mirror at all whilst I drive past the school.			

Bonus step: Checking out the daily checklist

Just to stress the importance of doing your ERP exercises every day and reviewing them, we've made a daily checklist to jog your memory and help you keep that commitment to your new goal (see the earlier section 'Step 1: Understanding the goal of CBT for OCD'):

✔ Did I do my ERP exercise today?

✔ Did I drop all my safety behaviours?

✔ Did I manage to perform no neutralising responses?

✔ Did I act in an anti-OCD manner?

If the answer to any of these questions is 'no', you need to think about what got in the way of performing the ERP exercises correctly. You may find the list of simple rules for successful ERP in Chapter 7 helpful.

When you can answer 'yes' to all the questions, you can either repeat the exercise until it becomes unproblematic or move on to another, more challenging ERP exercise.

Staying Motivated

Realistically, becoming super proficient at something is hard work, and it's even harder work when you have to put yourself through a considerable amount of discomfort (which may often feel more like excruciating pain and anguish). But you're not alone; people all over the world put themselves through hardship to reach their goals, whether they're Olympic swimmers getting up at 4 a.m. to train for three hours before school or a nomadic shepherd moving sheep from one side of the country to the other. Sticking to the plan and enduring the tough times is what gets people to their goals.

 To help keep yourself motivated, think of people who work hard to achieve their goals who inspire you – maybe a sportsman, family member or historical figure. When you feel yourself weakening, you can ask yourself what they'd have done at this point and what would've happened to their achievements if they'd stopped when the going got tough.

Being patient (Rome wasn't built in a day)

In most cases, making the changes you're after takes time and repetition. Everyone has a different pace of improvement, but it's still a case of taking one step at a time. Whether these are relatively small steps or bigger steps, continuing to put one foot in front of the other is what's going to get you to your goal. When you look at where you are now compared to where you want to be, you may find it overwhelming – rather like being at the bottom of a mountain, looking up at the peak and thinking you're never going to reach it. If your goal is to climb to the top of the mountain, there's only one solution, right? Keep going, and eventually you'll get there.

If you want an inspiring story about someone who took small steps to achieve what could easily have been seen as an impossible task, read or watch *Touching the Void*. It's the true story of a climber who had an accident near the top of a mountain and, against the odds, made a gruelling three-day journey back to base camp despite a broken leg, freezing temperatures and no food or water.

Recognising your progress

When you're still on the road to reaching your goal, you can easily feel that nothing is changing and you're never going to get there. Reaching a milestone is one thing; recognising how far you've come in getting to it is another. Remember that choosing to act differently in response to those thoughts, images, feelings and urges that are part of your OCD is a big step in itself. Whether you've come a long way or done a little to work toward reaching your goal, you need to be able to take stock and give yourself credit for what you've done. Take a moment every week to look back and see what has changed since you started this process.

The attitude you take with yourself about your progress (or lack of it) makes a difference to your motivation.

If you wouldn't use the things you say to yourself to motivate others, don't try to use them to motivate yourself! Think about what you'd say to someone if you wanted to help them recognise their progress on a long and difficult journey. Which would be more motivational: telling them they were going too slowly and were doing a rubbish job or encouraging them to focus on the progress they had made, however small? You don't respond to criticism and discouragement any differently to anyone else!

Rewarding yourself

The road to recovering from OCD can often be a long and rocky one, so it's important to reward yourself for both your achievements and your effort on a regular basis. Rewards are a way not only of marking your achievements and feeling good about them but also of incentivising yourself to tackle something you'd rather avoid. For this reason, we suggest that you think about how you're going to reward yourself in advance.

For example, you could tell yourself that after you've reached a milestone, you get to treat yourself to something you've been wanting.

Make a list of half a dozen things that you consider to be rewards. (Make sure these are realistic; for example, don't put 'buy myself a motorbike' as a reward if you're not going to be able to afford one.) They don't have to be big things. A reward may be lounging in the bath and having an hour to yourself. They just need to be achievable and enjoyable.

We suggest you reward yourself in two ways:

✔ **On a weekly basis (because you deserve it even if the progress is slow).** When you're planning your ERP at the beginning of the week (as we discuss earlier in the chapter), choose how and when you're going to reward yourself that week.

✔ **When you reach one of your milestones.**

We've created a simple chart for you to write down your milestones and the rewards you intend to give yourself. Note that it has a column for when you want to reach the milestones and one for when you actually reach them. However, the reward doesn't change whether you reach it early, on time or very late! The idea of setting a date is to give yourself something to aim for, but you don't have permission to beat yourself up or cancel the reward if you don't achieve it on time.

Milestone	Date I hope to reach it	Reward	Date reached	Reward redeemed?
To be able to write and send my Christmas cards without checking	November 21st	Play golf at the new course	December 14th	Yes, went with Tom

If you want to reward yourself with something big, then consider a reward jar. Instead of giving yourself a smaller reward, put marbles in the jar; when you have enough marbles, do or buy the reward for yourself.

Part IV
Move Over OCD – Putting Yourself in Charge

Myths versus Facts about Medication and OCD

Myth	Fact
Taking medication means I'm weak, and I should be able to do this on my own.	OCD is a serious problem and can be hard to overcome. There is no reason to feel more ashamed of taking medication for OCD than you would for a physical problem. OCD is an illness, not a test of your character.
OCD is either a chemical problem or a psychological one. If psychological treatment is available, medication shouldn't be necessary.	Spot the all-or-nothing thinking trap here. Your thinking and behaviour are products of your brain, and your brain is affected by the way you think and act.
If I start taking medication, I'll become addicted and end up having to take it for the rest of my life.	There is a small proportion of people who seem to do best staying on medication in the long term (who usually have a significantly better quality of life as a result). However, many others find that after they've improved, they can gradually reduce the dose in collaboration with their doctor or psychiatrist after a year or two.

Head to www.dummies.com/cheatsheet/ managingocdwithcbt for handy info and tips about managing OCD.

In this part. . .

- ✔ Focus on the values and activities you've shoved aside because of OCD.
- ✔ Look ahead to a future free from the control of OCD.

Chapter 9

Reclaiming Your Life from OCD

*A*s the saying goes, 'the wheel that squeaks the loudest gets the grease'. And with emotions like fear and guilt, plus strong urges to carry out compulsions, OCD can squeak so very loudly!

You may, like a lot of people, have had OCD for many years before seeking help. Even if you have had OCD for six months, your brain will start to carry out more compulsions and avoidance out of habit. As this happens over time, you can begin to feel like your OCD is part of *you*. Moreover, because it's like a parasite attaching itself to your moral code and desire to be safe or avoid causing harm, seeing where the real you stops and your OCD starts can become difficult.

Sadly, some people lose track of their true values, hopes and dreams – what they're really about as people – and start to think of their OCD as almost defining who they are. OCD is an illness (a clinical disorder recognised by healthcare professionals around the globe) and has no right to dominate your life.

Based on a CBT model called ACT (acceptance and commitment therapy) and the work of Steve Hayes and Kelly Wilson, this chapter is all about clarifying and reconnecting with the real you. Putting yourself in charge and not your OCD really helps you break free.

Valuing Values

Values are personal; only you can define what's important to you and what you want your life to be about. It's not up to us to tell you what's important to you, although most people usually share a limited number of values. Values aren't an all-or-nothing, achieve-it-or-don't concept. Rather, they're about making a commitment to small actions that are consistent with what you care about.

Values aren't goals with outcomes or things you can tick off a list. 'Making sure we spend at least one night out together each week' may be a goal in the pursuit of being a committed and supportive partner; 'training in CBT' is one goal in the pursuit of being a good therapist. Your values are the signposts pointing you in your important directions in life; your goal lies along those paths.

It doesn't matter whether your values are being a good parent or loving partner, helping animals, looking after the environment, helping others who need to be cared for or discovering knowledge. Some people emphasise one rather than another. What's important is that you take actions in ways that are consistent with what you care about.

So how do you clarify just what your values are? The following sections provide a few methods for distilling what's truly important to you.

Visualising the end of the line

One good (albeit macabre) way to examine your values is to envision looking back across your life from your deathbed. How do you think you'll need to have lived in order to feel that you've lived well? How well do your current avoidance, reassurance-seeking, ruminating, compulsions and so on fit in with living that way?

Similarly, think what you'd see and hear as a fly on the wall at your own funeral. What might people say about you? What would you want them to say; that is, what sort of person do you want to be remembered as? It's unlikely that you want to be remembered for the things your OCD makes you do or worry about. People aren't likely to celebrate the way you washed your hands, checked, monitored your bodily reactions or tried to ward off disaster by changing your thoughts.

Asking yourself questions about what's important

At times, seeing your way out of OCD through the forest of intolerance of uncertainty, fear, excessive responsibility and so on can be hard. Clarifying and committing to following the direction of your values can be your compass on this journey. Use the following questions to help clarify your values and remind yourself of what you're really about:

✔ **What's important to you in your romantic or intimate relationship?** What sort of partner do you want to be to your spouse or significant other? How has your OCD interfered with the way that you relate to your partner? Has your relationship become characterised by conversations about your worries, reassurance or implementing rules to try to avoid triggering your OCD? If you aren't involved in a relationship at present, how would you like to act in a relationship?

✔ **What's important to you in your family relationships?** As with romantic relationships, OCD can often influence and damage family relationships. How do you want to act as a sibling, child or parent/parent-in-law? If you aren't in contact with some of your family members, do you want to be? How would you act in such a relationship? How can you start to put this into action?

✔ **What sort of friend do you really want to be?** What's important to you in the way you act in the friendships you have? How do you want your friends to remember you? If you have no friends, would you like to have some? What role would you like in a friendship? How can you start to put this into action?

✔ **What's important to you as far as work is concerned?** What sort of employee or employer do you want to be? How important to you is what you achieve in your career? If you're self-employed or run a business, what sort of business do you want to run? If you're currently unable to work, what sort of values do you hold about work?

✔ **What's important to you regarding learning, education or training?** Learning and education may be a route to better grades, a better job, skill development or simply the pleasure and satisfaction of learning itself. What kind of learner or student do you want to be? What sorts of attitudes and practices do you want to embody? How can you start to put this into action?

✔ **What's important to you as far as hobbies or interests are concerned?** Hobbies and interests are good for your health, protecting against stress and depression and of course adding crucial variety to your week. What are your recreational activities? If you have none, what are some interests you want to pursue? How does your OCD interfere in this area? Do you feel you have no time for hobbies because your day is spent focused on feared catastrophes or moral dilemmas? How can you start to put this into action?

✔ **What's important to you regarding spirituality or religion?** If you're spiritual or religious, how do you want to follow that path? Studies show that practising a religion is good for your mental health. Does your OCD seriously interfere with your ability to engage in your religion, and, if so, does it cause significant distress? How can you reclaim your religious or spiritual practices?

✔ **How do you want to contribute to society or your community?** What's important to you in the way you build the sort of world that you want to live in? Are you interested in voluntary or charity work or political activity? It can be particularly important for some individuals with OCD to find a healthy outlet for their sense of responsibility to others. OCD tends to focus you on protecting people in very narrow, and usually unproductive, ways. What are your true values and desires as a member of society, and how can you express them in more rewarding and constructive ways?

✔ **What's important to you regarding your health and well-being?** Because OCD tends to inflate the importance of certain inner experiences (such as thoughts, images and so on) or narrow your focus to particular feared threats, you can easily end up neglecting your health. What's important to you as far as your mental health is concerned alongside your physical health? Do you regularly exercise, eat well and get an appropriate amount of sleep? If not, what can you do to improve in these areas?

Becoming More You, Less OCD

If you were starting your life from scratch as a designer or project manager, how much time would you plan to build in for OCD? Our guess is none at all. Clarifying your values and what you're truly about as a person isn't the answer in itself. However, given the force, energy, effort and persistence that you have to recruit at times to overcome your OCD, we hope that this exercise helps galvanise you in resisting being pushed around by your OCD.

Alongside the techniques outlined in this book, you can think of your recovery as crowding out the influence of your OCD and breaking free from its grip. Remember that there are only 52 weeks in the year, 7 days in each of those weeks and 24 hours in each of those days. Consider what portion of your time you want to set aside for each of your values and the activities that follow from them. Put the squeeze on your OCD by depriving it of time and energy as you put those resources into areas that really count. Then go in for the kill with your anti-OCD CBT.

Chapter 10

Building a Brighter Future

*A*s we discuss throughout this book, the cornerstone of recovering from OCD is to understand and fully accept that your thoughts, images, doubts and urges are entirely normal. To help your brain update and operate as if this is true, you need to show a clear commitment to training your brain into understanding the normality of these mental events.

In this chapter, we also consider some of the additional help that exists that may help you on your journey of creating a brighter future.

Refocusing Recovery

A full recovery doesn't in any way depend on having the mental events no longer enter your mind. A full recovery depends on reclaiming your mind, body, heart and soul from the clutches of your OCD. This chapter is all about rebuilding your life as you break free from OCD. We hope that this rebuilding is part of and flows from the results of standing up to your OCD.

It's really important not to see building your future free from OCD as a secondary concern. It's the whole point. It's also an integral part of how you keep your freedom from the tyranny of OCD. Think of your life as you would a garden. The various mechanisms that maintain your OCD are like weeds. You've been busily pulling them up, leaving some bare soil. Your next step will be to plant and tend your desired shrubs and flowers (healthy parts of you and your life).

Examining Neglected Parts of Your Personality

Consider the parts of your personality that may have become overdeveloped over the years and any parts that you may have neglected. Examples of personality traits that tend to be more highly developed in people with OCD are perfectionism and responsibility. Take a look at some of the personality traits that you may already have but have tended to neglect.

Creativity	Ingenuity	Curiosity
Open-mindedness	Good judgment	Spirituality
Love of learning	Perspective	Wisdom
Courage	Perseverance	Diligence
Industriousness	Honesty	Authenticity
Enthusiasm	Love	Connectedness
Kindness	Generosity	Social skills
Social intelligence	Teamwork	Fairness
Forgiveness	Humility	Prudence
Discretion	Self-control	Love of beauty
Love of excellence	Gratitude	Humour
Hope	Optimism	Playfulness

Taking Interest in Hobbies and Interests

Having OCD can partly reflect that you have an active brain that needs lots of stimulation. The truth is that most people are happier if they pursue absorbing hobbies, interests or education. Research shows that people who engage in hobbies and interests are significantly less likely to relapse in their OCD. Important stuff because if there is one thing that's almost as important as getting better, it's *staying* better.

 Grab a blank piece of paper and a pen and brainstorm any leisure activities, hobbies, learning, self-development or spiritual activities that you used to enjoy. Include ones you enjoyed as a child. Now add any activities that you've ever fancied giving a try. Step by step, think of activities you could carry out indoors, then outside; those you could do alone, then those that need one or more other people. If they're activities you think your OCD used to make difficult, all the better. Now choose about three or four to give a try over the next month and at least one you can carry out this week.

 There is no wrong answer as to what you choose to do with your time. The trick is to find activities that are a good match for you. No one can tell you what you like to do. If you're not sure, try out plenty and see.

Mastering Medication

Many people with OCD have been told (incorrectly) that medication is the only real treatment for OCD. That's not to say medication can't be very helpful for some people; taking medication isn't at odds with CBT. Some people liken medication to taking insulin if you're a diabetic, in that it's correcting for a deficit in the body. Scientists debate whether OCD has a neurological or biochemical cause, whether changes in the brain are in fact caused by having OCD, or whether medication for OCD simply helps but tells us little about the underlying cause of the problem (like taking a painkiller for a headache).

At the end of the day, what counts is your recovery. Taking medication is a decision that you of course need to make with your general practitioner or psychiatrist. Our contribution is to help debunk some of the less helpful ideas that float around about taking medication, as shown in Table 10-1.

Table 10-1   Myths versus Facts about Medication and OCD

Myth	Fact
Taking medication means I'm weak, and I should be able to do this on my own.	OCD is a serious problem and can be hard to overcome. There is no reason to feel more ashamed of taking medication for OCD than you would for a physical problem. OCD is an illness, not a test of your character.
OCD is either a chemical problem or a psychological one. If psychological treatment is available, medication shouldn't be necessary.	Spot the all-or-nothing thinking trap here. Your thinking and behaviour are products of your brain, and your brain is affected by the way you think and act.
Medication may have harmful side effects, and taking it is irresponsible.	Having OCD has immense effects on your life. There is evidence that it leads to neglect of physical health in some individuals. In response to the stress and strain of OCD, your body produces its own chemical responses. Having OCD isn't a natural or healthy state.
If I start taking medication, I'll become addicted and end up having to take it for the rest of my life.	There is a small proportion of people who seem to do best staying on medication in the long term (who usually have a significantly better quality of life as a result). However, many others find that after they've improved, they can gradually reduce the dose in collaboration with their doctor or psychiatrist after a year or two.
Taking medication now means I've failed. It's like admitting defeat.	This point of view largely stems from the rather outdated way in which society views mental health problems. Somehow, mental health is seen as within human control and physical health outside human control. Both assumptions are wildly wrong.
	Having OCD is not your fault. You can't just snap yourself out of it. You should feel no worse about taking medication to help your fight against OCD than you would about using an anti-inflammatory if you were doing physiotherapy for a back problem. If you've found it hard to defeat OCD without medication and you're being encouraged to try antidepressants, think of accepting them as adding another weapon to your armoury. You're still the one fighting the battle.
I've tried medication before, and it just doesn't work for me.	People often have to try more than one antidepressant. Having the correct dose for OCD is also important (it's usually considerably higher than that for depression).

Considering Professional Help

The most important thing to remember is that psychological help for OCD needs to be specifically tailored for the problem. Some people get a bit bogged down in seeking a therapist who specialises in their particular type of OCD. This focus may be problematic in the sense that it narrows down the range of possible sources of help dramatically and may mean that you're putting too much emphasis on the content of your obsessions rather than the processes (such as avoidance and compulsions) that maintain them. Check that the therapist is appropriately qualified and experienced in treating OCD.

Collaborate with your therapist. This isn't an exercise in shifting the burden of responsibility or seeing whether you're worth caring about. Here are the sorts of things you and your therapist can work on together:

✔ See using CBT for OCD as an experiment. Put your own (and possibly other people's) doubts and reservations about recovery to the test. Make a fresh start.

✔ Know your enemy. Get a good general understanding of the attitudes and maintaining mechanisms of OCD and how they maintain the problem. Write this information out in a list, a picture, a diagram – any format that helps you keep the information clear.

✔ Identify the main anti-OCD strategies you're going to try out.

✔ Start to keep track of your avoidance, frequency of mental and physical compulsions and so on.

✔ Consider involving key individuals who may be participating with your OCD (such as giving reassurance) and work with them to guide them on standing up to your OCD.

✔ Commit to anti-OCD rehabilitation. Deliberately practise confronting triggers with new mental and physical responses.

✔ Reclaim your life.

✔ Look after yourself well as you fight back and overcome your OCD.

✔ Build your free life and work on relapse prevention.

Part V

The Part of Tens

For ten bonus tips on coping with OCD recovery, head to
www.dummies.com/extras/managingocdwithcbt.

In this part. . .

- ✔ Help your loved one battle OCD.
- ✔ Develop certain traits and strategies to help overcome OCD and keep it at bay.
- ✔ Keep an eye out for common OCD obstacles.

Chapter 11

Ten Tips for Family and Friends of a Person with OCD

* *

In This Chapter

▶ Supporting your loved one, not his OCD

▶ Helping yourself weather the storm

* *

*W*hen someone close to you is suffering from OCD, you're likely to feel an impact from it. This chapter is about ways to deal with that impact constructively. Seeing someone you care for struggle is difficult, and you likely want to try to help your loved one feel better. Unfortunately, well-meaning friends and family often unwittingly give support that helps maintain the problem rather than lessen it. You may have engaged in some of the behaviours we talk about in this chapter. If that's the case, don't give yourself a hard time! You didn't know any better and were only doing your best to help. We talk you through the common pitfalls and discuss the alternatives so that you're better equipped to help both your loved one and yourself.

Remembering That Your Loved One Isn't His OCD

OCD can be a relentless, infuriating, stubborn and cunning beast, so your ability to make a distinction between your loved one's OCD and your loved one himself is really important.

In treating OCD, we encourage individuals to see their OCD as a bully that is trying to control them. This approach helps people take a step back and see the OCD as something outside themselves, which in turn can help them choose not to give in to the bully's demands. For the same reason, you also need to be able to distinguish between these two.

You may have found yourself wondering what's happened to the reasonable person you used to know. That person is still there but is just overshadowed by the OCD bully. When your loved one is acting in ways that are out of character and seem totally unreasonable, then you're most likely faced with the OCD bully's demands rather than those of your loved one. As hard as it can be, bearing in mind that your loved one isn't being deliberately difficult can help you to remain calmer and more compassionate to him (but not to his OCD). Your loved one needs your support more than ever, but you need to discover how to support the person rather than the OCD.

Demonise the OCD and not your loved one.

Realising You Can't Force Someone to Change

Seeing clearly what someone's problem is and how to fix it is only part of the battle; the other person has to be willing to admit there's a problem and seek help.

The willingness of the OCD sufferer to commit to change and engage in the therapeutic process is a big factor in getting a positive outcome from therapy. In other words, forcing someone who is resistant to try to change or go to therapy may well be a waste of time and effort (and often money). So as counter-intuitive as it may seem, you may need to continue to provide gentle encouragement and suggestions regarding seeking help instead of delivering an ultimatum. You may even find that the harder you push, the more resistance you encounter.

This resistance doesn't mean, however, that you can do nothing to help. Apart from giving the person support, you can stop unintentionally supporting the OCD, as we describe in this chapter. The more you're able to resist accommodating the OCD, the more difficult it will be for the OCD's roots to grow stronger. By refusing to give in to the OCD's demands, you may well make life more difficult for the person with OCD in the short term; that may seem unkind, but when you stop making things easier, the person is more likely to consider changing or seeking help. Ultimately, helping the person to recover is the only truly kind thing to do.

Avoiding Giving Reassurance

As you're probably aware, OCD is a disorder that is typically characterised by distress and doubt. One important part of the recovery process is about learning to deal with that doubt differently – namely, feeling the doubt and giving up trying to get certainty. Seeking reassurance is a compulsion, which the OCD sufferer needs to learn to give up to overcome the problem. You can help your loved one stop asking for reassurance by figuring out how to stop giving it.

If you've ever been asked for reassurance from an OCD sufferer (for example, 'Are you sure I didn't say something offensive?' or 'Do you think that's safe?'), you've most likely discovered that giving reassurance doesn't make the doubt go away for very long. It may disappear for a few moments or even hours, but inevitably the doubt returns. OCD demands 100 percent certainty (99.99 percent won't do), and because being 100 percent certain is impossible, reassurance can never work long term.

 As we mention throughout the book, OCD is very sneaky, so you need to get well acquainted with the ways in which your loved one seeks reassurance:

- ✔ **Directly:** Asking whether something happened or whether something is safe, asking you to check something, calling to ask whether you're okay

- ✔ **Indirectly:** Relating something to you to gain reassurance from your reaction, calling about something banal but using the call to check that you're okay, asking about 'normal' behaviour

When you notice you're saying things like 'It's fine', 'Stop worrying', 'It'll be okay' and 'Yes, I'm sure', you know you're giving reassurance. You can take two useful avenues to replace giving reassurance:

  ✔ Have a discussion with the person suffering from OCD to explain that you're aware that providing reassurance is unhelpful and that you're no longer going to do it. When the person then asks for reassurance, you can simply let him know that he's doing it and that you aren't going to engage in it; for example, 'It's just your OCD trying to control you; don't respond, and it will pass'.

  ✔ Train yourself to give answers that leave plenty of room for doubt, such as 'I don't know', 'I'm not sure', 'You'll have to wait and see', 'Maybe' and so on.

When we talk about reassurance here, we're talking very specifically about trying to get rid of the person with OCD's obsessive doubts. We aren't suggesting you withdraw emotional support. So it's okay (within reason) to reassure your loved one that you're there for him and that you'll help him get through this.

Letting the Person with OCD Set the Pace

Becoming well acquainted with how OCD works (as you're doing now) is a great idea so that you're best placed to provide help and support. However, remember to allow your loved one to take the journey at his own pace. Trying to push the person along can be tempting, but pushing too hard before someone is ready can be counter-productive and worsen rather than alleviate the OCD.

However much you've acquainted yourself with OCD, don't try to act as a therapist if you aren't specifically trained as one.

That said, giving encouragement and making suggestions based on a shared understanding of the problem can be useful. (Just be prepared to let go if your recommendations aren't welcomed.) If you believe the individual with OCD could be doing more and is giving in to the OCD, you may want to help him think about this by asking him the following:

- ✔ Can you think of a helpful anti-OCD behaviour?
- ✔ Are you being driven by yourself or your OCD?
- ✔ Has this behaviour worked for you in the past?
- ✔ Is there something you can do to work toward your goal?
- ✔ Is there anything constructive I can do to help you disengage from the grip of the OCD bully (help you engage in something else, such as listening to music, going for a walk together and so on)?

Refusing to Take Part in Rituals

Rituals, like reassurance seeking, are compulsive behaviours that compound the problem of OCD rather than make it better. Many friends and family of OCD sufferers end up engaging in rituals for several reasons:

- ✔ It seems like the right thing to do.
- ✔ The rituals help relieve anxiety on a short-term basis.
- ✔ Agreeing to engage in rituals is often quicker or the path of least resistance.
- ✔ The person with OCD begs them to help or has a compelling argument as to why they should.

 The simplest way to think of rituals' effects is to compare them to scratching an insect bite. It relieves the discomfort in the short term but makes the bite worse longer term because the more the spot gets scratched, the more it needs scratching.

For someone with OCD, the desire to escape from uncomfortable feelings is very strong, which is why people engage in rituals to provide immediate relief even though they don't help long term. Learning to endure the short-term discomfort is vital to overcoming OCD. Therefore, the people around a person with OCD need to demonstrate the 'short-term pain, longterm gain' principle, even when the person in front of you is stomping, crying, pleading and anxious. Whenever you choose to take part in any kind of ritual, you're colluding with the OCD and suggesting that short-term relief outweighs long-term benefit. This shift can be painful for you in the short term; seeing someone in distress is unpleasant.

So what to do instead? If you're already engaging in certain rituals with your loved one, we aren't suggesting you go cold turkey. A better idea is to discuss these rituals with your loved one and agree a plan for slowly stopping engaging in them. This change may lead the other person to move to other rituals that appear even worse to you; don't let this switch put you off! For any new rituals that you're asked to engage in, simply decline (ignoring the foot stomping, pleading and so on as much as possible) and explain that by taking part in the rituals you're helping to make the problem worse in the long term.

When you're no longer part of the rituals, not only is life better for you, but the ball is also entirely in the person with OCD's court. The more rituals he has to perform himself, the harder it becomes; the harder it becomes, the more likely the person is to want to tackle the OCD constructively.

You need to be strong to withdraw from or refuse to engage in rituals. However, if you don't demonstrate the 'short-term pain, long-term gain' principle, you can't very well expect somebody with OCD to do it!

Recognising Progress, However Small

Overcoming OCD is no mean feat and is certainly not something that happens overnight. This wait can be tough when you're rooting for someone you desperately want to get better. One of the difficult parts of supporting someone trying to overcome OCD is that improvements are often invisible to the non-sufferer. Particularly at the beginning of treatment, improvement is measured by looking at the intensity, frequency and duration of obsessions and the accompanying feelings. So someone with OCD may note a less frequent desire to respond to intrusive thoughts and the ability to curb the unhelpful response a little more quickly – things the onlooker can't see. Instead, the onlooker often just sees that the person is still reacting to the intrusive thoughts and draws the conclusion that nothing is changing.

We encourage people with OCD to be patient in making progress, and we strongly recommend that you follow suit. Pushing for faster change is likely to demotivate, whereas supporting and rewarding even small changes can give the individual with OCD encouragement to carry on.

Understanding that progress may be slow and that sometimes change isn't even visible to you should help you not to get too frustrated with the process. Make sure that you're also doing your part in not unwittingly supporting the OCD because being less entangled with the OCD means you can more easily remain supportive rather than feel frustrated or resentful. (See 'Avoiding Giving Reassurance' and 'Refusing to Take Part in Rituals' earlier in this chapter for info on disentangling from another's OCD.)

Engaging in More Interesting Pursuits

Life needs to be about more than OCD, both for you and your loved one. OCD loves a vacuum, so it gets worse the more you're both kicking around dealing with the OCD or waiting for it to strike rather than getting out there and attempting to do interesting, fun things.

Think about what you've stopped doing since you've been busy dealing with your loved one's OCD and the fallout from it. You need to be engaging in the things that make you happy and give you fulfilment now more than ever. Taking time for yourself doesn't make you uncaring; on the contrary, feeling content makes you better able to support and assist. When you go out and live your life, you're being a good role model for your loved one with OCD.

Refraining from Being Accommodating

Even strong, healthy relationships come with an element of compromise and the need to accommodate certain behaviours in each other. When you have a relationship with an OCD sufferer, the compromise often ends up being rather one-way, and you may have found yourself accommodating all sorts of things you'd never have dreamed of.

Common forms of accommodation include the following:

✔ Avoiding doing things that may trigger the OCD

✔ Modifying work, social or family routines

✔ Assisting with simple, everyday tasks

✔ Assuming the other's responsibilities

✔ Tolerating abnormal or extreme behaviours

✔ Putting up with unpleasant conditions

✔ Participating in rituals

✔ Repeatedly giving reassurance

Understand that in doing these things, you're accommodating the OCD rather than your loved one. These demands and conditions come directly from the OCD bully.

Though accommodating the OCD's demands can be tempting and is often the easiest short-term solution, it doesn't work in the long term, for many reasons:

✔ It gives the OCD attention it doesn't deserve.

✔ It encourages the OCD bully to keep making more outrageous demands.

✔ It gives the person with OCD the message that responding to the OCD in this way is okay or necessary.

✔ It supports the belief that the OCD sufferer can't cope or needs to be protected.

✔ Tolerating difficult or unusual circumstances can increase frustration and resentment in those affected.

✔ It makes the problem worse in the long term.

So what's the next move if you've turned your life upside down accommodating the OCD? You need to gradually return your routine and lifestyle to normal. This process may not be easy and may encounter the person with OCD's resistance, so try to create a shared understanding of why you're making these changes and agree a plan for withdrawing from accommodating the OCD.

Sometimes you just can't reach an agreement with the individual who has OCD, perhaps because he isn't ready to change or is still too much under the influence of the OCD bully to be able to see the merits. In this case, you may choose to wait and revisit the conversation at a later date; however, be warned that you may never get an agreement and that waiting for the green light isn't a prerequisite to withdrawing from accommodating the OCD bully. If your life has become a struggle because of your accommodating your loved one's OCD, you may need to make these changes regardless of whether the other person likes it.

When you've decided on a plan for withdrawing from accommodating the OCD (with or without agreement), be clear with your loved one and yourself that it's a case of *when* you stop accommodating the OCD and not if. A person who knows you mean business is more likely to take it seriously and get on board with the idea.

You need to be strong and consistent and stick with your plan. If you refuse to accommodate one day but give in the next, the OCD will just keep trying to get you to give in, rather like an unruly toddler.

Recognising OCD Disguised as a Legitimate Concern

As we reiterate throughout this chapter, you most effectively help your loved one by responding differently to the OCD bully and not giving in to its demands. So the question becomes, how do you tell the difference between OCD and other worries? The simplest thing is to assume that all worries, concerns and anxiety are as a result of the OCD and treat them as such. Ninety-nine percent of the time, that is indeed the case.

If you want to get clarity about whether a concern is related to OCD, simply ask how the person would feel if you didn't help him at that moment. If the answer is that he'd feel anxious (or any other emotion that typically comes with his OCD), then it's OCD. If he says he'd not feel anxious, suggest that he test this theory out to see what happens. He may refuse, which suggests that the concern is in fact OCD-related; it was just trying to disguise itself to get you to do what it wanted!

Although we encourage you to be mindful that many of your loved one's worries will be related to OCD, this does not mean that everything you find difficult or annoying about them will necessarily be OCD-related. Sometimes they may be putting up a fight simply because they want something different to you. Bear in mind how frustrating it would be if every time you did something someone else didn't like they wrote it off and refused to discuss it. If you have an open discussion about the fears and worries that are targets for change, you may be better able to notice whether something is OCD-related.

Responding When Nothing Seems to Be Changing

When you've done your best to understand the problem and tried to support your loved one in beating OCD but nothing is changing, you understandably may feel angry, impatient or demoralised. Here are a few tips to help you:

- ✓ **Troubleshoot.** Have you stopped accommodating the OCD as best you can?

- ✓ **Don't take it personally.** Remember that you can do your part, but in the end the person with OCD is responsible for choosing to get better.

- ✓ **Don't judge.** Nobody enjoys having OCD; it's not easy to overcome, so assume the person is doing the best he can in the current circumstances.

- ✓ **Accept that you can't force someone to change.**

- ✓ **Stick to your part even if nothing appears to be changing.** If you stick to not accommodating the OCD, you do help in the long term.

- ✓ **Review your options.** If your loved one is in therapy but isn't improving, you may want to discuss the reason and whether other options are available.

- ✓ **Shift your focus.** If you can't promote change in your loved one's OCD, let it be for now and focus on creating fulfilment for yourself.

Chapter 12

Ten Things You Need More of to Help Fight Your OCD

● ●

In This Chapter

▶ Embracing the feelings your OCD wants you to avoid

▶ Remembering that fun, laughter and activities are all great tools

▶ Recognising the importance of taking care of yourself

● ●

*S*aying that living with OCD isn't much fun would be a gross understatement. OCD is anything but fun. When you're struggling with OCD, you mainly spend your time and energy trying to avoid or minimise distress; this demanding schedule not only maintains the problem but also robs you of the opportunity of enjoying yourself. Trying to keep yourself as certain and risk-free as possible denies you possibilities and minimises your ability to be spontaneous, follow your dreams and live your life the way you want.

This chapter is all about trying to get more of the good, helpful stuff rather than focusing on trying to get rid of the bad, unhelpful stuff. We encourage you to take a double-pronged attack by squeezing the OCD from both sides: by behaving in an anti-OCD manner to reduce the power of the problem and by engaging in more life-enhancing activities and behaviours to give the OCD less space to breathe, boosting your mood and strengthening your resolve.

Risk Taking

OCD makes people very risk averse, and the idea of engaging in more risk taking seems somewhat perverse. However, the more risks you take, the more you can start to live outside of the confines of your OCD. Taking risks may feel deeply uncomfortable (see the later section 'Willingness to Experience Distress and Discomfort'), but it becomes easier the more you do it. Without risks, everyday life is pretty much impossible. Risks are everywhere: risks that you'll be late for work, that someone will steal your washing if you hang it out on the line, that you'll be hit by a freak storm, that you'll trip and hurt yourself, that your loved one will be in an accident and so on. There are so many risks that they could fill this book on their own.

Living is one big risk. Anything can happen, and as much as you may want to control and minimise risk, so many things are out of your control that spending all that time and energy trying to minimise risk is simply futile. Working to control the uncontrollable is exhausting. So instead of trying to take fewer risks (you know it's not helping), throw the cat among the pigeons and take more risks. These risks don't have to be solely about your OCD. Think about other areas of your life where you've stopped taking risks, whether it's at work, at school or in your personal life.

Make a list of areas where you want to experiment with taking more risks and then choose some risks to take. You may be pleasantly surprised with the results!

Tolerance of Uncertainty

When you have OCD you tend to crave certainty, but the more you try to get it, the less certain you feel in the long run. A good example of this concept is the contestants on quiz shows where big prizes are at stake. Often, the person clearly knows the answer but starts to second-guess or doubt herself because she's trying to be absolutely sure that she's giving the right answer (so as not to lose the prize). You can bet your bottom dollar that if she were at home, she'd be shouting at the TV with plenty of conviction; the only difference would be that because she had nothing at stake, she wouldn't be demanding certainty and therefore would feel *more* certain.

No doubt before you started trying to tackle your OCD differently you spent lots of time and effort trying to avoid uncertainty and to feel absolutely sure about things. However, now you know that this pursuit is counter-productive and can see uncertainty as your new best friend!

Remind yourself that the more uncertain you feel (without responding or trying to resolve anything), the more you're helping yourself learn not to need certainty. The less you demand certainty, the more certain you will feel.

Humour

As Oscar Wilde said, 'life is much too important a thing ever to talk seriously about it', which is a reminder that taking things too seriously makes them harder rather than easier. When you lose your sense of humour, everything feels like a struggle, whether it's something as small as untying a knot or as big as losing a job. Even when loved ones die, family and friends gather together and get comfort from reminiscing over funny moments, using smiles and laughter to help them through the pain.

So observe your daily life and ask yourself whether there's enough humour in it. A good dose of the giggles improves your mood no end, so even if you can't laugh at your OCD right now, find things that do make you laugh, such as watching a favourite comedian or TV show, hanging out with particular friends who make you laugh or doing something 'silly' – whatever works for you.

 Because one of the biggest maintaining factors for OCD is taking your thoughts, feelings and urges too seriously, test out what happens when you recognise the ridiculousness of the thoughts and laugh at them instead.

Good Quality Sleep

Don't underestimate the importance of a healthy sleep routine. Sleep is the time your brain and body repair themselves and process things. Studies recommend that adults get seven to eight hours of sleep a night; although this mark isn't always possible, you should aim for something around it. Putting up a fight against the OCD is even harder when you're tired, so do yourself a favour and get some more sleep. Next time your OCD is trying to get you to stay up late to do compulsions, tell it to get lost! You can always do your rituals in the morning if you can't stop yourself, but putting them off often diminishes the urge to do them.

Set yourself a daily wake-up time and stick to it regardless of what time you went to bed. If you wake up at 8 a.m. despite having gone to bed at 3 a.m., you're less likely to stay up the following night engaging in unhelpful behaviours (whereas sleeping in gives more room for the OCD to keep you up the following night). Sticking to this plan helps you resist the urge to engage in compulsions when you should be in bed.

In our experience, most people with OCD get too little sleep rather than too much; however, if you're in the camp of getting too much sleep, cut back to the recommended amount (give or take) and see what happens. Oversleeping can actually make you more tired (weird but true), and it's almost certainly a way of avoiding getting on with your life, which maintains the OCD.

Spontaneity

Increasing spontaneity is a great way of employing a host of anti-OCD weapons at once: It challenges you to take risks, promotes feeling uncertain, helps you relinquish control, leaves room for imperfection and supports your living the life you want rather than being beholden to the OCD's demands.

Try doing more things off the cuff, last minute or unplanned. Often, people assume that they need to maintain control and order to feel in power and remain on an even keel, but remarkably, the opposite is often the case.

Throw caution to the wind and see what happens. Notice whether it was as difficult as you anticipated and observe how you coped despite the lack of planning.

Healthy Eating

We're not dieticians and aren't going to advise you on exactly what to eat, but we do want to remind you that having a healthy diet and eating regularly impacts your mood and your OCD more than you imagine. When you look after your body by putting good fuel in it (think fruit, vegetables, fibre and protein) and reduce the amount of unhealthy fuel you take on board (like sugar, alcohol and fast food), you benefit both your body and your mind. Looking after yourself by eating healthily is a way of valuing yourself.

Often, OCD affects what you think you can eat; if this is the case for you, that's even more reason to choose how and what you want to eat and stick with it despite the OCD.

Willingness to Experience Distress and Discomfort

The very mention of feeling more distress frequently inspires confusion or even anger, but bear with us and all will become clear. As you know only too well, when you have OCD you experience plenty of distress and discomfort despite all your efforts to avoid or minimise them. Think of this as *wasted distress*, in that despite feeling it, you're not getting any closer to overcoming your problem. Now imagine choosing to challenge your OCD to bring on and practise tolerating the distress and discomfort. Think of this approach as *worthwhile distress* because the more you learn to tolerate it, the closer you get to your goal of overcoming your OCD. Either way you feel distress, so you may as well feel the worthwhile kind!

Change your attitude toward distress and discomfort and see these feelings as tools to work with to overcome your problem rather than obstacles to be avoided.

Distress and discomfort are your new best friends: The more you deliberately bring these feelings on and tolerate them, the better you get.

Hobbies and Activities

OCD tends to take up a whole lot of time; for this reason, OCD sufferers often feel they no longer have time to engage in the activities they enjoy. OCD expands to fill the time available to it, so the more you focus on doing the things you enjoy (rather than what your OCD wants you to do), the less time you have for engaging in OCD rituals and ruminations.

When you do things you enjoy, be they sewing, skydiving or whatever, you remind yourself of what it means to be you and can more easily stand up to the OCD bully.

Consider committing to doing an enjoyable activity that involves other people. Doing so helps you stick to your plans. When the only person who loses out if you bail is you, OCD can much more easily persuade you to do its thing rather than yours. But if you're letting someone else down by changing your plans, you're more likely to fight the OCD bully.

Acceptance

Acceptance is often confused with resignation, but it's anything but. *Acceptance* is about recognising things for the way they are so that you can choose to think and behave in the manner most helpful to you.

In dealing with OCD, acceptance is particularly important in several areas:

- ✔ **Acceptance of OCD:** Again, this concept isn't about resigning yourself to OCD or to liking it in any way, shape or form. It's about choosing to accept it as a problem that needs dealing with. Often, people doubt that what they're experiencing is actually OCD and worry that it may be something more sinister. When you accept that it's OCD or choose to treat it 'as if' you believe it's OCD, you can practise responding in helpful ways.

- ✔ **Acceptance of intrusions:** When you accept that you can't force yourself not to have intrusive thoughts, images, doubts or sensations however hard you try, you free yourself from maintaining the vicious cycle and fighting a losing battle.

- ✔ **Acceptance of misconceptions:** This point is about how other people see OCD and how they respond to it. Many people simply don't understand what OCD is or how it works, and they offer you unhelpful advice like 'try not to worry about it' or 'just be strong and don't do it'. These well-meaning platitudes can be frustrating and anger-provoking, but if you accept that often people simply don't get it, you free yourself from an unrealised expectation and can focus on helping yourself rather than on worrying about what others think.

✓ **Acceptance of yourself:** Perhaps the most important thing of all is to be able to accept yourself despite the issues you're facing. Having OCD doesn't mean anything about who you are. Remind yourself that you're a fallible and complex human being (just like everyone else on the planet) who can think and feel bad things and good things.

Exercise

People with OCD commonly compromise on the simple elements of living a healthy life. Looking after yourself creates a strong, healthy platform from which you're better able to handle your OCD (and anything else that life throws at you).

Exercise not only makes you physically healthier but also impacts your mood by releasing *endorphins*, the chemicals that make you feel more energised and happier.

You don't have to become a gym bunny to fulfil the requirement of getting enough exercise. Simply choosing to walk to work or take the stairs can make a difference. If you can combine getting exercise with doing something enjoyable, like having a walk in the park or playing a team sport, then so much the better.

Exercise doesn't have to be boring. Think of an activity that gets butterflies going in your tummy or makes you feel like a child again, such as going on a swing or a trampoline. Try things out: The aim is to ultimately find something that's a good enough fit for you to keep doing it!

Chapter 13

Ten Traps to Avoid in Recovering from OCD

In This Chapter

▶ Falling prey to fear and uncertainty

▶ Having unrealistic expectations

*Y*our OCD won't go down without a fight. Your brain is quick to maintain habits (no matter how unhelpful!) and the very thinking that causes OCD is likely to get in the way of recovery. The best way to avoid any obstacle is to see it coming. To help with this, here are some of the more common obstacles we've seen people encounter over the years.

Needing to be 100 Percent Sure That You Have OCD before You Get Started

Even getting a diagnosis of OCD from a professional doesn't prevent people from questioning whether they're 100 percent sure they have OCD. The trap is part of the 'allergic' attitude toward uncertainty that underpins OCD. Don't wait until you're absolutely certain it's OCD before making changes; you'll never be absolutely certain. If the symptoms you have are a good enough fit for the diagnosis of OCD, you'll learn lots more by seeing how they respond if you treat your condition as if it's OCD.

Fearing That Change Will Be Too Difficult

Breaking free from OCD can be difficult, although many people tell us that it's not much more difficult than living with OCD. Actually, the real issue here is the gap between the anticipated difficulty of standing up to your OCD (which may or may not be as bad as you think) and the promised (but at best rarely delivered) relief of getting to the place where your avoidance, compulsions and other safety-seeking strategies have worked. OCD is tough. Being free from it is better. Change is hard but very well worth it. Try it out. You can always go back to your original strategies if you discover they were truly working better. Don't let OCD bully you out of your right to give change a go.

Mistaking Asserting Your Rights with the Tyranny of OCD

For some people, this trap can be very important. They become convinced that people should respect their right to avoid triggers or to have their rules and rituals complied with. This mindset can lead to a lot of friction with others and leave the person with OCD putting time and effort into arguments that are utterly self-defeating. What these folks forget is that their OCD and not their true selves steers their drive for their 'rights'. Take care to pursue your values and passions, but make sure they aren't your OCD in sheep's clothing.

Looking for Just the Right Person to Help You

Getting the correct treatment for your OCD is important, but notice whether looking for the right therapist or psychiatrist is hindering or helping. If you find your professional's not all that understanding, supportive or constructive, then by all means ask to see someone else. But be carful not to let your OCD manoeuvre you into avoiding seeing any professional because that person may not be *the* professional for you.

Insisting on a 100 Percent Thorough Treatment

A common compulsion in OCD is the need to be fully understood, which can sometimes lead to some very long, repetitive conversations and exhausted (and frequently none-the-wiser) therapists. Some people also get bogged down with the idea that they need to get to the root of it all and get it all sorted out. There is in fact some debate as to just how much human beings can be completely figured out from their early childhood issues to their adult bad habits. Don't lose the forest for the trees. OCD is so big and ugly that improving it imperfectly is likely to have a very substantial impact on your quality of life.

Being Unclear on Your Goal

Oddly enough, overcoming OCD can't be your main goal. Your brain has a hard time maintaining an absence of something. The key is to see overcoming OCD as a means to an end. Think carefully about how you want to be different as a result – something to really visualise and aim for. Use this goal to give you a focus when you're working hard to beat your obsessions, face your fears and stop your rituals. Head to Chapter 8 for more on clarifying your OCD goal.

Confusing Freedom from OCD with Freedom from Intrusive Thoughts

As we note throughout the book, you have no control over what pops into your mind. Intrusive thoughts, doubts, images and urges are completely normal. They're not OCD, and they're not the problem. It's the fear, guilt, shame and other forms of discomfort that are the issue (as are the amount of time consumed and the rituals and avoidance that you may engage in). Work on improving your behaviours and emotional response and keep the purity of allowing your mental events to take care of themselves, and you find a much clearer path to recovery.

Being Too Afraid of Making Things Worse

No one wants to make a problem like OCD worse. However, one of the most common reasons a problem becomes chronic is the fear that an attempted change will make things worse. Therapy is an experiment that may make you feel worse in the short term but is very much more likely to help you get better in the long run. This fear is entirely natural and best treated as another 'what-if' to practise on.

Going for 'Normal' Too Soon

Most people, to a large degree, just want to be 'normal'. The problem is that real fitness and good health rarely come from normal. Physical healthcare is very often rather abnormal. Physiotherapy exercises can seem positively weird, and medications and operations are often far outside a person's everyday experiences. Getting truly fit means a level of training that is almost certainly not the norm.

The point, of course, is that treatment for OCD is no different. The most important element is the extent to which you have to deliberately seek out your feared/avoided triggers to get sufficient practice instead of relying on the more normal-seeming approach of cutting back on rituals such as checking.

Not Filling the Void Early Enough as a Relapse Prevention Measure

As soon as you start driving your OCD out of your life, begin to install other important activities in its place. You often hear experts say that OCD loves a vacuum or OCD makes work for idle hands (or minds). When you cut a consuming activity like maintaining OCD out of your life, you create such a vacuum. Give your mind help to stay away from the behavioural and mental processes of OCD by building a life that includes activities that fill the void.

Index

About the Authors

Katie d'Ath is a CBT therapist who specialises in treating anxiety and obsessive-compulsive spectrum disorders. She has an MSc from Goldsmiths College, University of London and is accredited by the British Association of Behavioural and Cognitive Psychotherapists. Before setting up in private practice, she worked in the National Health Service in the UK and at The Priory Hospital North London.

Katie is particularly interested in raising awareness of mental health problems and improving access to self-help. She has a YouTube channel dedicated to providing people with free CBT tutorials. She has made several TV appearances, regularly presents at mental health charity conferences and is on the clinical advisory board for OCD Action. This is her first book.

Rob Willson is a CBT therapist based in North London, with a special interest in obsessional problems. He currently divides the majority of his work time between seeing patients, conducting research, writing and teaching.

He is the chair of the Body Dysmorphic Disorder (BDD) Foundation, the world's first charity exclusively devoted to BDD. Prior to building his own practice, Rob spent 12 years working at the Priory Hospital North London, where he was a therapist and therapy services manager. He also trained and supervised numerous CBT therapists over a seven-year period working at Goldsmiths College, University of London and completed his PhD at the Institute of Psychiatry in London.

Rob has co-authored several books, including the bestselling *Cognitive Behavioural Therapy For Dummies* (Wiley) and *Overcoming Obsessive Compulsive Disorder*. His main clinical interests are anxiety and obsessional problems and disseminating CBT principles through self-help. He has been featured in several newspaper and magazine articles and has made several TV and radio appearances.

Dedication

We dedicate this book to all of the volunteers, staff and professionals involved in the charities around the world that provide
information and support for people affected by OCD.

Authors' Acknowledgments

We'd very much like to acknowledge the individuals with OCD, their families, partners and friends who have been the inspiration for this book. OCD can be a truly devastating problem, and it is the courage, determination, kindness and humour so often shown by people with OCD that makes the job of being a therapist so rewarding.

We'd also like to thank the others who helped bring this book into fruition. In particular, Annie Knight and Christina Guthrie, who have been hugely supportive, flexible and patient. We also very much appreciate the technical review of the manuscript and encouraging comments from our colleague Nick Page and copy editing by Megan Knoll. We are, as always, grateful to our partners and families, who supported us in numerous ways throughout this process.

Further acknowledgment is due to all of the clinicians and researchers who have contributed so much to the field of behaviour therapy and cognitive behaviour therapy for OCD, without whom this book very definitely would not exist. Amongst others, these include David A. Clark, Lynne Drummond, Edna Foa, Mark Freeston, Isaac Marks, Stanley Rachman, Adam Radomsky, Paul Salkovskis, Roz Shaffran, Gail Steketee, David Veale and Adrian Wells.

Publisher's Acknowledgments

Executive Commissioning Editor:
Annie Knight

Editorial Project Manager:
Christina Guthrie

Development Editor: Christina Guthrie

Copy Editor: Megan Knoll

Production Editor: Siddique Shaik

Cover Image: ©iStock.com/maxuser

Take Dummies with you everywhere you go!

Whether you're excited about e-books, want more from the web, must have your mobile apps, or swept up in social media, Dummies makes everything easier.

Visit Us

Like Us

Follow Us

Watch Us

Join Us

Pin Us

Circle Us

Shop Us

FOR DUMMIES®

A Wiley Brand

BUSINESS

978-1-118-73077-5

978-1-118-44349-1

978-1-119-97527-4

MUSIC

978-1-119-94276-4

978-0-470-97799-6

978-0-470-49644-2

DIGITAL PHOTOGRAPHY

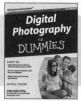

978-1-118-09203-3

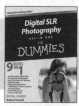

978-0-470-76878-5

978-1-118-00472-2

Algebra I For Dummies
978-0-470-55964-2

Anatomy & Physiology
For Dummies, 2nd Edition
978-0-470-92326-9

Asperger's Syndrome For Dummies
978-0-470-66087-4

Basic Maths For Dummies
978-1-119-97452-9

Body Language For Dummies,
2nd Edition
978-1-119-95351-7

Bookkeeping For Dummies,
3rd Edition
978-1-118-34689-1

British Sign Language For Dummies
978-0-470-69477-0

Cricket for Dummies, 2nd Edition
978-1-118-48032-8

Currency Trading For Dummies,
2nd Edition
978-1-118-01851-4

Cycling For Dummies
978-1-118-36435-2

Diabetes For Dummies, 3rd Edition
978-0-470-97711-8

eBay For Dummies, 3rd Edition
978-1-119-94122-4

Electronics For Dummies
All-in-One For Dummies
978-1-118-58973-1

English Grammar For Dummies
978-0-470-05752-0

French For Dummies, 2nd Edition
978-1-118-00464-7

Guitar For Dummies, 3rd Edition
978-1-118-11554-1

IBS For Dummies
978-0-470-51737-6

Keeping Chickens For Dummies
978-1-119-99417-6

Knitting For Dummies, 3rd Edition
978-1-118-66151-2

FOR DUMMIES

A Wiley Brand

SELF-HELP

978-0-470-66541-1

978-1-119-99264-6

978-0-470-66086-7

LANGUAGES

978-0-470-68815-1

978-1-119-97959-3

978-0-470-69477-0

HISTORY

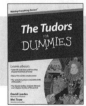

978-0-470-68792-5

978-0-470-74783-4

978-0-470-97819-1

Laptops For Dummies 5th Edition
978-1-118-11533-6

**Management For Dummies,
2nd Edition**
978-0-470-97769-9

Nutrition For Dummies, 2nd Edition
978-0-470-97276-2

Office 2013 For Dummies
978-1-118-49715-9

Organic Gardening For Dummies
978-1-119-97706-3

Origami Kit For Dummies
978-0-470-75857-1

**Overcoming Depression
For Dummies**
978-0-470-69430-5

Physics I For Dummies
978-0-470-90324-7

Project Management For Dummies
978-0-470-71119-4

Psychology Statistics For Dummies
978-1-119-95287-9

**Renting Out Your Property
For Dummies, 3rd Edition**
978-1-119-97640-0

**Rugby Union For Dummies,
3rd Edition**
978-1-119-99092-5

Stargazing For Dummies
978-1-118-41156-8

**Teaching English as a Foreign
Language For Dummies**
978-0-470-74576-2

Time Management For Dummies
978-0-470-77765-7

Training Your Brain For Dummies
978-0-470-97449-0

**Voice and Speaking Skills
For Dummies**
978-1-119-94512-3

Wedding Planning For Dummies
978-1-118-69951-5

WordPress For Dummies, 5th Edition
978-1-118-38318-6

Think you can't learn it in a day? Think again!

The *In a Day* e-book series from *For Dummies* gives you quick and easy access to learn a new skill, brush up on a hobby, or enhance your personal or professional life — all in a day. Easy!

Available as PDF, eMobi and Kindle